Lotus Illustrated

DICTIONARY of CHEMICAL ENGINEERING

Compiled and Edited by:
Tiffany Watson

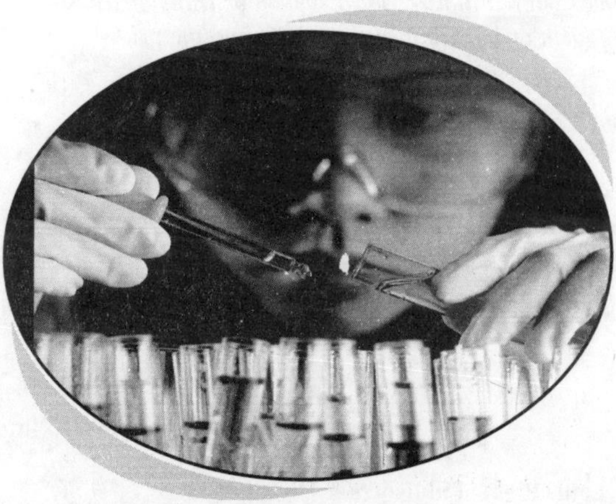

Lotus PRESS
4735/22,Prakash Deep Building
Ansari Road, Darya Ganj,
New Delhi-110002

Lotus *Illustrated*
DICTIONARY *of*
CHEMICAL ENGINEERING

ISBN 81 89093 20 7

Published by:
Lotus Press : Publishers & Distributors
Unit No. 220, 2nd Floor, 4735/22,
Prakash Deep Building, Ansari Road, Darya Ganj,
New Delhi- 110002, Ph.: 32903912, 23280047, 09811838000
• E-mail : lotus_press@sify.com, www.lotuspress.co.in

Printed at : ***Afcons Enterprises,*** New Delhi

Preface

Chemical engineering is one of the most important and fundamental branches of engineering. All matter has a chemical constituent. Whether it is computers, electronics, textile, and electrical engineering or medicine, one has to work at chemical level. Study of chemical part of the things helps in better understanding of the nature of that matter. This further helps in simplification of the task of proper usage of that material.

For those involved in chemical studies, good knowledge of specific terms is very essential. And this is not the end of the story. As there are many elements, and all of them have their specific chemical and physical properties, it becomes a gargantuan task to remember these details. On the other hand, when many elements combine to form something different, as happens in chemical engineering, the task becomes more complex and difficult. To overcome this state of affairs one has to have access to an exhaustive and reliable dictionary.

This dictionary is an attempt to give an access to the whole paraphernalia of relevant terms in a crisp package. The terms have been explained in simple and comprehensible manner, with relevant pictures and illustrations to bring clarity. The dictionary can be taken as a complete and cognisable set of chemical engineering terms to be used as a referral. Since it is a vast subject, a complete inventory of the terms can hardly be claimed in one lexicon, but nonetheless, an attempt has been made towards providing a gamut of terms, which form a part and parcel of the subject.

■ **a/d converter**

a to d means analog to digital. This electronic hardware converts an analog signal like voltage, electric current, temperature, or pressure into a digital number that a computer can process and interpret.

■ **100% acrylic latex paint**

water-thinned paint in which only acrylic resin is used as the binder medium. Typically the highest quality latex paints used for a wide variety of architectural coatings, 100% Acrylic Latexes have superior adhesion, long-term flexibility, breathability, alkali resistance, toughness, and colour and sheen retention.

■ **2-bromo-2 nitropropane 1-3 diol**

an antimicrobial agent, active against bacteria and fungi. It can react with secondary amines and form nitrosamines, which are known carcinogens. Used at high levels it can cause irritation.

■ **5-hydroxyindoleacetic acid**

the acid found as a cerebral spinal fluid in the body, thought to be responsible for the degree of aggressive behaviour in humans.

■ **abel tester**

a closed-cup flash tester for kerosene and other oils.

■ **abortifacient**

induces the premature abortion of the foetus. Example: pennyroyal, aloe, sandalwood.

■ **abrasion**

1. the wearing away or cleaning by friction. Abrasion can also relate to the wearing away of a floor finish film by friction.
2. abrasive.
3. a hard and wear-resistant material that is used to wear, grind or cut away other material.

■ **abrasive**

1. a product that works by abrasion. Products such as cleaners, polishes and pads may contain an abrasive.
2. absolute.
3. an aromatic, volatile substance obtained by solvent extraction from a single botanical species - for example rose absolute and jasmine absolute. None of the solvent should remain after the process is complete when absolutes may be used like essential oils.

■ **absolute pressure**

in a region such as outer space, which is virtually void of gases,

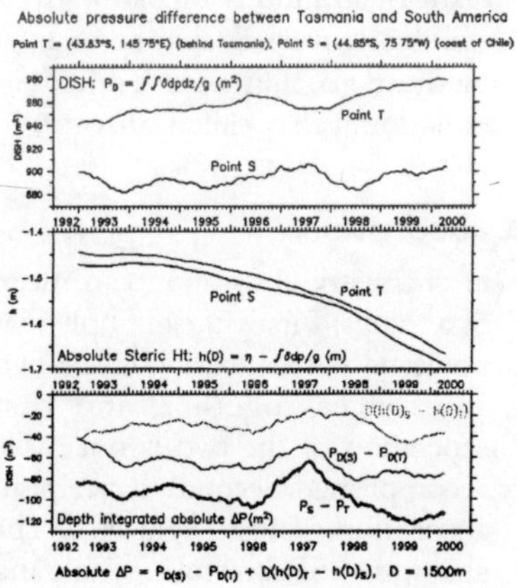

the pressure is essentially zero. Such a condition can be approached very nearly in a laboratory when a vacuum pump is used to evacuate a bottle. The pressure in a vacuum is called absolute zero, and all pressures referenced with respect to this zero pressure are termed absolute pressures.

■ absolute risk

is the excess risk due to an exposure to a hazard.

■ absorbance

absorbance (*A*) is defined as - log(1-*á*) = log(1/*ô*), where *á* is the absorptance and *ô* the transmittance of a medium through which a light beam passes.

■ absorbed dose

for any ionising radiation, absorbed dose (*D*) is the mean energy imparted to an element of irradiated matter divided by the mass of that element.

■ absorptance

absorptance (*á*) is ratio of the radiant or luminous flux in a given spectral interval absorbed in a medium to that of the incident radiation. Also called absorption factor.

■ absorption

in chemistry absorption can mean two things: Firstly it can imply that powerful forces exist holding two substances together, and that seperation of the two is not easily accomplished. Secondly it can mean absorption of heat, light etc... The absorption of ultraviolet, visible and infrared radiation is the basis of some forms of spectrometry which can be used to identify different chemical compounds.

2) Absorption

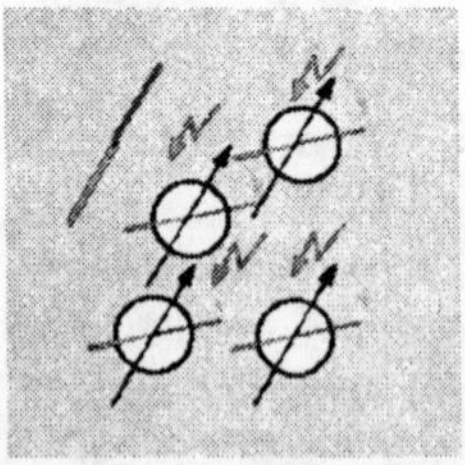

■ absorption coefficient

absorption coefficient (*a*) is the relative decrease in the intensity of a collimated beam of electromagnetic radiation, as a result of absorption by a medium, during traversal of an infinitesimal layer of the medium, divided by the length traversed.

■ absorption coefficient, molar

molar absorption coefficient (*å*) is absorption coefficient divided by amount-of-substance concentration of the absorbing material in the sample solution (å = a/c). The SI unit is m^2mol^{-1}. Also called extinction coefficient, but usually in units of $dm^3cm^{-1}mol^1$.

■ abstracted model

model that reproduces or bounds the essential elements of a more detailed process model and captures uncertainty and variability in what is often, but not always, a simplified or idealised form. see **abstraction**.

abstraction

the essential components of a process model that are extracted for use in a total system performance assessment. The abstraction retains the basic intrinsic form of the process model but does not usually require its original complexity.

acceptable risk

the concept of acceptable risk is not particular easy to define. It is essentially a measure of the risk of harm, injury or disease arising from a chemical or process that will be tolerated by a person or group.

accessible environment

any point outside of the controlled area, including 1. the atmosphere (including the atmosphere above the surface area of the controlled area), 2. land surfaces, 3. surface waters, 4. oceans, and 5. the lithosphere.

accumulation

accumulation is the gradual build-up of a chemical as a result of repeated doses or exposure, where destruction or elimination of the chemical fails to keep pace with its introduction into the body. This will result in a gradual increase in concentration in an organism, organ or tissue which may cause toxic effects that can become more marked with successive doses. Factors that can enhance accumulation include selective binding of the chemical to particular types of molecules in the body, the progressive concentration of fat-soluble chemicals in body fat, complete failure or a slowing of metabolism which would otherwise reduce the concentration of chemicals, and slow excretion.

acerola

a cherry from Bermuda, high in natural Vitamin C, a powerful antioxidant.

acetamide mea

a synthetic raw material used as a hair conditioner and solvent. It has low toxicity to the skin and eyes.

acetic acid

an acid in vinegar and used as a preservative in foods and a solvent for gums and volatile oils.

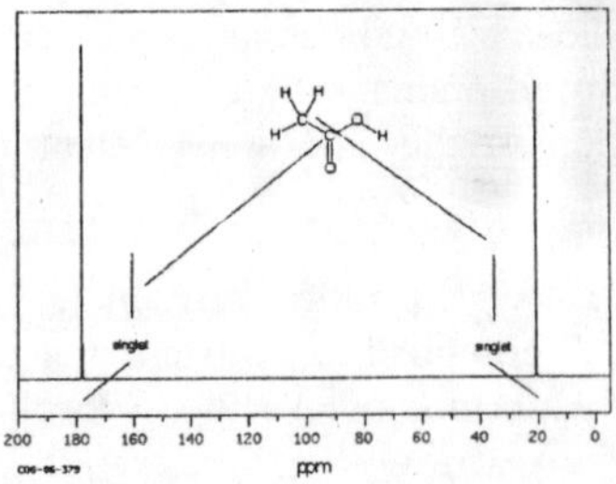

acetone

a solvent used mainly in nail varnishes and nail polish removers. It is an irritant, and its use in cosmetics has been banned or restricted in many countries.

acid

1. acid is a type of compound that contains hydrogen and dissociates in water to produce positive hydrogen ions.

2. a substance that when dissolved in water dissociates and can do-

nate a hydrogen (proton) to another molecule. Examples include sulphuric acid and nitric acid.
3. a compound that ionises in water to produce hydrogen ions. It readily donates protons to other substances and, when dissolved in water, creates solutions that conduct electricity, taste sour and turns litmus paper red. Inorganic acids (sometimes called mineral acids) include sulfuric, nitric, hydrochloric and phosphoric. Organic acids include acetic, oxalic, hydroxyacetic and citric. Acids are used in toilet bowl cleaners, rust removers and hard water stain removers.

■ acid colours

synthetic dyes used to colour food, cosmetics and toiletries. The list of allowable acid colours has been reduced dramatically in recent years because they are highly irritant.

■ acid dissociation constant

acid dissociation constant (K_a) is the equilibrium constant for the dissociation of an acid HA through the reaction

$HA + H_2O$ à$A^- + H_3O^+$

The quantity $pK_a = -\log K_a$ is often used to express the acid dissociation constant.

$pK_a = -\log K_a$

Acid Dissociation Constant *examples*

1. Write down the expression for K_a for each of the following acids:

(a) CH_3COOH

$CH_3COOH + H_2O \longrightarrow CH_3COO^- + H_3O^+$

(b) H_2CO_3

$H_2CO_3 + H_2O \rightleftharpoons HCO_3^- + H_3O^+$

(c) HCO_3^-

$HCO_3^- + H_2O \rightleftharpoons CO_3^{2-} + H_3O^+$

■ acidity

the presence of acid-type constituents whose concentration is usually defined in terms of neutralisation number. The constituents vary in nature and may or may not markedly influence the behaviour of the oil.

■ acidophiles

a classification of organisms suited to survive and flourish in regions of extremely high acid content.

■ acrylic

a synthetic polymer used in high-performance latex or water-based paints. As the paint's binder, acrylic resins enable the coating to last longer and retain its colour.

■ acrylic latex paint

water-thinned paint which employs acrylic resin as the majority of the binder. Other binders which may be added to reduce cost or add specific properties include styrene, epoxy, and poly-vinyl acetate.

■ acrylic resin

resins which have established a pre-eminent position among coating formulators, having shown superiority in such respects as colour and gloss retention, alkali and oxidation (chalk) resistance, hardness, adhesive and cohesive strength, and overall film durability. Generically, resins result from the polymerisation of derivatives of acrylicacids, including esters of acrylic acid, methacrylic acid, acrylonitrile, and their copolymers. Also known as acrylate resins.

actinides

this is a group of elements within the periodic table and are also known as the transuranics. The name refers to those elements with increasing atomic no. from actinium, which all have similar chemical properties, like the lanthanides. The first few members of the group are the naturally occuring elements actinium, thorium, proactinium and uranium. Beyond this elements have been made artificially by radioactive bombardment. These artificial elements are unstable, some have very short lifeimes and undergo spontaneous radioactive decay.

activation energy

constant E_a in the exponential part of the Arrhenius rate equation associated with the minimum energy difference between the reactants and an activated complex, (transition state which has a structure intermediate to those of the reactants and the products) or with the minimum collision energy between molecules that is required to enable a reaction.

active ingredients

the ingredients contained in any formula that give the desired physiological effect - for example, the component in a moisturising cream that improves the moisture content of the skin.

activity

1. activity (a) is a thermodynamic function used in place of concentration in equilibrium constants for reactions involving non-ideal gases and solutions. For the species i activity is defined as

$a_i = f_i \times c_i$

where a_i is the activity of the species i, c_i is its molar concentration, and f_i is a dimensionless quantity called the activity coefficient.

2. in adsorption chromatography, the relative strength of the surface of the packing. For silica gel, the more exposed the silanol groups, the more active the surface. Activity can be controlled by adding water or another polar modifier, which is hydrogenbonded to the active sites, thereby reducing the surface activity.

activity coefficient

activity coefficient ($\tilde{a}$ or f) is a fractional number which when multiplied by the molar concentration of a substance in solution yields the chemical activity. This term gives an idea of how much interaction exists between molecules at higher concentration. In solutions of very low ionic strength, when m is less than 0.01, the Debye-Hückel limiting law can be used to calculate approximate activity coefficients where $\tilde{a}_i$ = activity coefficient of the species i, z_i = charge on the species i and $\imath$ = ionic strength of the solution.

acute effect

an acute effect is one which involves severe symptoms which develop rapidly and may quickly reach a crisis.

■ acute hazard

an acute hazard is one to which a single exposure may cause harm, but which is unlikely to lead to permanent damage.

■ additive effect

an additive effect is one in which the effect of two chemicals acting simultaneously is the simple sum of the effects that they would have if acting alone.

■ adhesion

the ability of dry paint to remain on the surface without blistering, flaking or cracking. Adhesion is probably the single most important property of paint. Wet adhesion, the ability of dry paint to adhere to the surface in spite of wet conditions, is particularly important for exterior house paints.

■ adhesive

any compound that can stick two surfaces together is classified as an adhesive. Simple adhesives are of cellulose, starch and rubbers. Modern adhesives are based upon complex polymeric materials. The adhesive is spread in an unpolymerised form, and the adhesive properties increase as polymerisation occurs between the two surfaces.

■ adiabatic

a chemical or mechanical process which takes place without heat entering or leaving the system. The term is only applicable to enclosed and isolated systems - so in essence it is idealistic and purely theoretical and is important in the study of thermodynamics.

■ adiabatic calorimeter

a calorimeter in which reactions proceed with little exchange of heat between the sample container and its surroundings. Common types of adiabatic calorimeter include the ARCTM, PHI-TECTM and VSPTM. The RSSTTM calorimeter may be considered as pseudo-adiabatic under some circumstances.

■ adiabatic induction time

induction period or time to an event (spontaneous ignition, explosion, etc.) under adiabatic conditions starting from specified operating conditions.

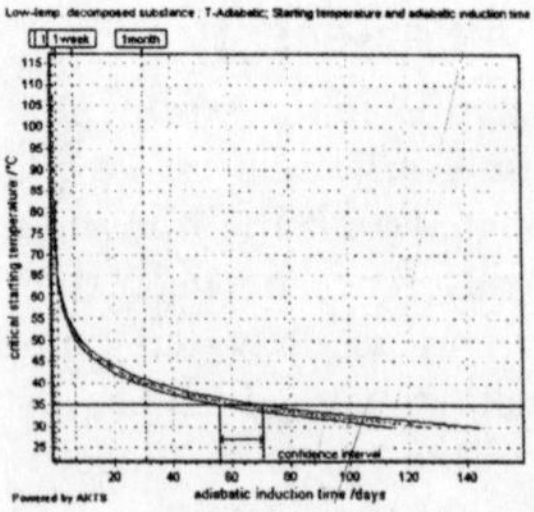

■ adiabatic process

adiabatic process is a thermodynamic process in which no heat enters or leaves the system. In general, an adiabatic change involves a fall or rise in temperature of the system.

■ adiabatic temperature rise

the increase in temperature of a reacting mixture as a result of exothermic chemical reaction, when there is no heat transfer to or from the environment. T = H/Cp where T is the adiabatic temperature rise (K), H is the heat release (J/kg) and Cp is the mean heat capacity (J/kg/K).

■ adjusted retention time

(t'_R) a measure of the retention time adjusted for the void volume. $t'_R = t_R - t_m$ (or t_0).

■ adjusted retention volume

(V'_R). Adjusts the retention volume tor the dead volume. $V'_R = V_R - V_m$, where V_m, is the dead, or void, volume.

■ adrenaline

this was the first naturally produced hormone to be isolated it a pure state. It is known as *epinephrine*, but its chemical name is *1-[3,4-dihydroxyphenol]-2-methylaminoethanol*. Its main action is to raise blood pressure, producing a faster pulse rate.

■ adsorbent

packing used in adsorption chromatography. Silica gel and alumina are the most frequently used adsorbents in HPLC.

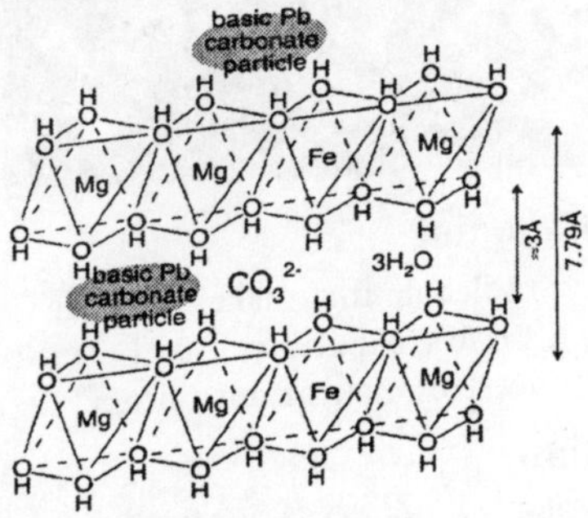

■ adsorption

1. the process of interaction between the solute and the surface of an adsorbent. The forces involved can be strong (for example, hydrogen bonds) or weak (van der Waals forces). For silica gel, the silanol group is the driving force for adsorption, and any solute functional group that can interact with this group can be retained by liquid-solid chromatography on silica.

2. the attachment of atoms, ions, or molecules of a gas, liquid, or dissolved substance to a surface. The term *sorption* is sometimes used interchangeably with this term.

3. if any compound, solid, liquid or gas, is loosely held by weak attraction to the surface of a solid it is said to have undergone adsorption. This process is much weaker and less permanent than absorption.

■ adsorption chromatography

one of the basic LC modes. Relies on the adsorption process to effect the separation. Silica gel and alumina are the most frequently used supports; molecules are retained by the interaction of their polar functional groups with the surface functional groups (e.g., silanols of silica).

■ adsorption isotherm

in adsorption, a plot of the equilibrium concentration of sample in the mobile phase per unit volume vs. the concentration in the stationary phase per unit weight. The shape of the adsorption isotherm can determine the chromatographic behaviour of the solute tailing. fronting, overload, etc.

Gas	Percent (%)
Nitrogen	78.008
Oxygen	20.95
Argon	0.93
Carbon Dioxide	0.03
Neon	0.018
Helium	0.0005
Krypton	0.0001
Xenon	0.00005

■ air set cement

a cement that sets through loss of water.

■ air-fuel ratio

the ratio of air weight to fuel weight consumed in an internal combustion engine or furnace.

■ airless spraying

process of atomisation of paint by forcing it through an orifice at high pressure. The effect is often aided by the vapourisation of the solvents, especially if the paint has been previously heated.

■ alara

abbreviation for as low as is reasonably achievable A process that applies a graded approach to reducing dose levels to workers and the public and reducing releases of radioactive materials to the environment. The goal of this process, often referred to as ALARA, is not merely to reduce doses, but to reduce them to levels that are as low as reasonably achievable.

■ albedo

a measure of reflectivity

■ alchemy

several of the great Greek philosophers considered all matter to be composed of four basic '*elements*', *fire*, *air*, *earth* and *water*, and that all materials were had these components mixed in different proportions. If this theory was true, then it was believed that all substances could be converted (*transmuted*) into each other by varying the proportions of the mixture. This tansmutation theory was the basis of Alchemy from approx. 300 B.C. to 1500 A.D., after which it slowly gave way to the more scientific concepts of chemistry from the 17th century onwards. The alchemists represented their '*element*' by a variety of symbols (see below), some of which were based on astrological signs

They were mainly concerned with trying to turn base metals into gold and looking for the elixir of life although they also developed medicinal drugs, developed couterfeiting techniques and the debasement of precious metals.

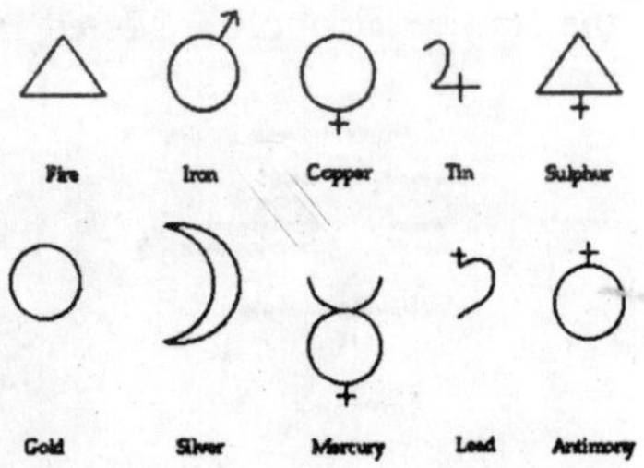

■ alcogel

colloidal gel made with alcosol.

alcohol

1. ethanol - olourless, volatile, flammable liquid produced by the fermentation of yeast and carbohydrates and used as a highly effective preservative. Can cause dryness and irritation when used topically. Ethanol can increase the permeability of the skin and may enable unwanted impurities to be absorbed. Alcohol is used frequently as a solvent and is also found in beverages and medicine. As an ingredient in ingestible products, alcohol may cause body tissues to be more vulnerable to carcinogens.

Mouthwashes with an alcohol content of 25% or more have been implicated in mouth, tongue and throat cancers.

2. an organic compound which has the general formula $\mathbf{C_nH_{2n+1}OH}$, they consist of hydrocarbon chains terminated by hydroxyl groups, **O-H**. Smaller members are water soluble, flammable and are useful as organic solvents and fuels. As with hydrocarbons, each member differs from the previous by an additional $\mathbf{-CH_2-}$ group. Alcohols with branced chains are also possible. Some of the simpler alcohols are listed below:

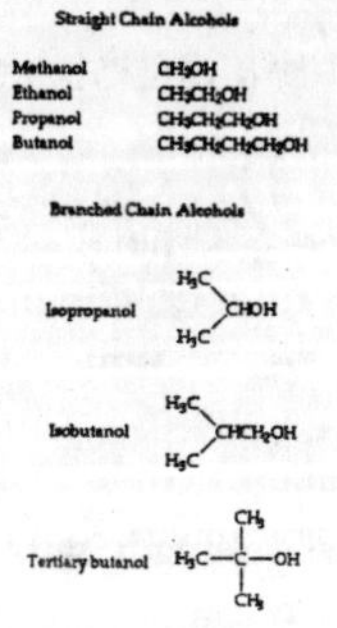

alcosol

a solution made with a mixture of a colloid and an alcohol.

alcove

a small excavation (room) off a main drift. Used for scientific study or for installing equipment.

aldehydes

aldehydes are broad class of organic compounds having the generic formula RCHO, and characterised by unsaturated carbonyl group (C=O). They are formed from alcohols by either dehydrogenation or oxidation. Their chemical derivation is indicated by the name *al*(cohol) + *dehyd*(rogenation).

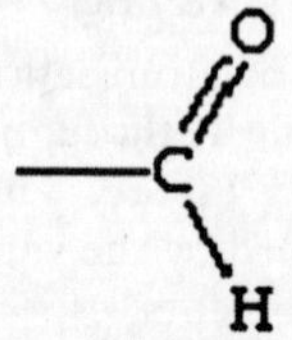

alicyclic compounds

an alicyclic compound contains rings of $\mathbf{-CH_2-}$ units joined by single bonds. Their general formula is $\mathbf{C_nH_{2n}}$, the simplest member being *cyclopropane*. An example of these compounds is *cyclohexane* (see below):

AC

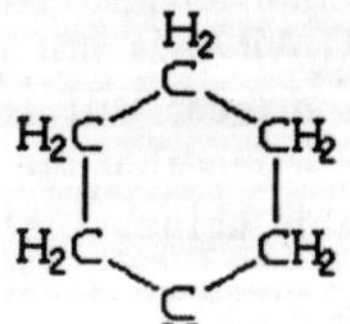

■ **alignment marks**

images selectively placed within or without an array for either testing, aligning or both.

■ **alignment**

the proper relative position accuracy of a graphic image, on an emulsion photo resist film mask with respect to an existing graphic image on a substrate.

■ **aliphatic compounds**

aliphatic compounds are acyclic or cyclic, saturated or unsaturated carbon compounds, excluding aromatic compounds.

■ **aliphatic solvents**

these are sometimes referred to as paraffins. They are also referred to as straight chain or open chain solvents. Kerosene, odourless mineral spirits and mineral seal oil are examples of aliphatic solvents.

■ **aliphatic**

a class of saturated or unsaturated carbon compounds, in which the carbon atoms are joined in open chains.

■ **alkali**

a water soluble hydroxide on one of the alkali metals. The term is virually synonymous with the term base. An example is *Sodium Hydroxide*, **NaOH**.

■ **alkali metals**

alkali metal is a term that refers to six elements: lithium (Li), sodium (Na), potassium (K), rubidium (Rb), caesium (Cs), and francium (Fr). These elements make up group 1 of the periodic table of elements. They all form singly charged positive ions, and are extremely reactive. They react violently with water, forming hydroxides and releasing hydrogen gas and heat. Caesium and francium are the most reactive and lithium is the least.

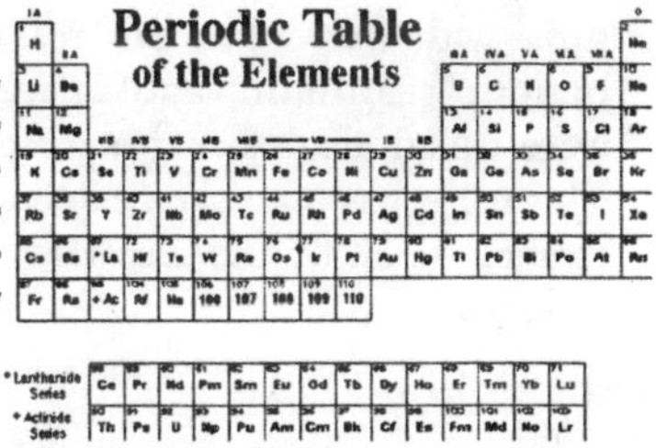

■ **alkali or base**

describes a solution formed when a base dissolves in water to form a solution which contains more hydroxide ions than hydrogen ions. Alkaline solutions have a pH of more than 7, turn red litmus paper blue, and feel soapy because they react with the skin. Alkalinity is exhibited in solution by alkalies such as sodium or potassium hydroxide or alkaline salts such as sodium carbonate. A substance used in some wax strippers, degreasers and cleaners to assist in soil and finish removal.

■ **alkaline earth metals**

alkali earth metal is a term that refers to six elements: beryllium (Be), magnesium (Mg), calcium (Ca), strontium (Sr), barium (Ba), and radium (Ra). These elements make up group 2 of the periodic table of elements. They all exhibit a single oxidation state, +2. They are all light and very reactive. Barium and radium are the most reactive and beryllium is the least. Chemists to denote slightly soluble metal oxides formerly used the term 'earth'. The oxides of barium, strontium, and calcium resemble alumina (Al_2O_3), a typical 'earth', but form alkaline mixtures with water. For this reason barium, strontium, and calcium were called alkaline earth metals. This name has now been extended to include all of the elements of group 2.

■ alkaloid

a naturally occuring organic compound containing nitrogen that acts as a base. Many alkaloids are physiologically active and can be used in small quantities as medicines, but if taken in larger doses they can be extremely poisonous. An example is caffeine.

■ alkalophiles

organisms that flourish in an environment where the pH is extremely high.

■ alkanes

alkanes (paraffins) are acyclic branched or unbranched hydrocarbons having the general formula C_nH_{2n+2}, and therefore consisting entirely of hydrogen atoms and saturated carbon atoms. In systematic chemical nomenclature alkane names end in the suffix -ane. They form a homologous series (the alkane series) methane (CH_4), ethane (C_2H_6), propane (C_3H_8), butane (C_4H_{10}), etc. The lower members of the series are gases; the high-molecular mass alkanes are waxy solid. Generaly the alkanes are fairly unreactive. They form haloalkanes with halogens when irradiated with ultraviolet radiation. Alkanes are present in natural gas and petroleum.

Some simple alkanes are listed below:

Straight Chain Alkanes

Methane	CH_4
Ethane	CH_3CH_3
Propane	$CH_3CH_2CH_3$
Butane	$CH_3CH_2CH_2CH_3$

Branched Chain Alkanes

Isobutane

Isopentane

Neopentane

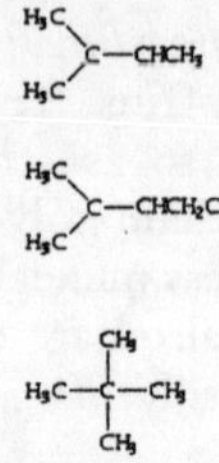

■ Alkenes

alkenes are acyclic branched or unbranched hydrocarbons having one or more double carbon-carbon bonds in their molecules. In systematic chemical nomenclature, alkene names end in the suffix -ene. The general formula is $C_nH_{(2n+2)-2x}$ were x is the number of double bonds. Alkenes that have only one double bond form a homologous series: ethene (ethylene), $CH_2{=}CH_2$, propene, $CH_3CH_2{=}CH_2$, etc. Alkenes typically undergo addition reactions to the double bond.

■ alkoxysilane

a reactant used for the preparation of chemically bonded phases. It will react with silica gel as follows:

R_3SiOR+ SiOH àSi-O-SiR_3 + ROH

■ alkyds

resins used mostly in trim paints, inside and out, although some medium duty equipment and marine enamels employ these resins as binders. Most often alkyd resins are found in vehicles employing aliphatic hydrocarbons (mineral spirits or other refined petroleum distillate) as thinner.

■ alkyloamides

a common family of ingredients, including Cocamide DEA, MEA, MIPA or PEG, used in shampoos, bubble baths and liquid hand and body cleansers. They are employed for thickening, gelling, emulsifying and solubilising. The major drawback with alkyloamides is that they can be contaminated with nitrosamines during manufacture of the product. Nitrosamines are suspected animal carcinogens. Correct formulation and the use of a reputable supplier will help eliminate the risk.

■ alkynes

1. alkynes (acetylenes) are acyclic branched or unbranched hydrocarbons having a one or more triple carbon-carbon bond. In systematic chemical nomenclature alkine names end in the suffix -yne. The general formula is $C_nH_{(2n+2)-4x}$ were x is the number of triple bonds. Alkynes that have only one triple bond form a homologous series: ethyne (acetylene), CH=CH, propyne, CH_3CH=CH, etc. Like alkenes, alkynes undergo addition reaction.

2. these compounds are also similar to alkanes. They have the general formula $\mathbf{C_2H_{2n-2}}$ corresponding to carbon chains with a triple carbon-carbon bond included. The simplest member of this family is ethyne (also known as *acetylene*). The triple bond makes these compounds very reactive. Some simple alkynes are listed below:

Simple Alkynes

Ethyne	$H—C{\equiv}C—H$
Propyne	$H—C{\equiv}C—CH_3$
1-Butyne	$H—C{\equiv}C—C_2H_5$
2-Butyne	$H_3C—C{\equiv}C—CH_3$

■ allantoin

derived from comfrey root - humectant, anti-irritating, healing and soothing.

■ allele

different forms of a gene that may be found at the same locus on a chromosome

■ allergic contact dermatitis

allergic contact dermatitis is a type of skin hypersensitivity. Its onset may be delayed by several days to as much as several years, for weaker sensitisers. Once sensitised, fresh exposure to the sensitising material can trigger itching and dermatitis within a few hours.

■ allotropy

1. allotropy is the occurrence of an element in two or more crystalline forms.
2. certain chemical elements have the ability to exist in two or more different structural forms known as allotropes. These allotropes may possess different physical properties such as density and melting points. Allotropic elements include carbon, tin, phosphorus and sulphur. Each allotrope is stable within a certain range of temperature and pressure only, and under certain conditions an allotrope can be converted into another.

■ alloy

1. a metallic solid or liquid formed from an intimate combination of two or more elements.
2. an alloy consists of an intimate mixture of two elements, usually metals to give a metal compound or solid solution. *Bronze* is an alloy of copper and tin and *brass* on of copper and zinc. By varying the composition it is possible to alter the chemical and physical properties of the alloy such as conductivity, ductility and corrosion resistance.

■ alloy 22

a corrosion-resistant, high-nickel alloy used for the outer barrier of the waste package.

■ aloe barbadensis

aloe Vera - soothing and cooling. An anti-irritant with healing properties.

■ alpha hydroxy acid

an organic acid produced by anaerobic respiration. Skin care products containing AHA exfoliate not only damage skin cells, but the skin's protective barrier as well. Long term skin damage may result from its use.

■ alpha particle

alpha particle is a helium nucleus emitted spontaneously from radioactive elements both natural and manufactured. Its energy is in range 4-8 MeV and is dissipated in a very short path, i.e. a few centimetres of air or less than 0.005 mm of aluminium. As a helium nucleus consists of two protons and two neutrons bound together as a stable entity the loss of an alpha particle involves a decrease in nucleon number of 4 and decrease of 2 in the atomic number. A stream of alpha particles is known as an alpha ray or alpha-radiation.

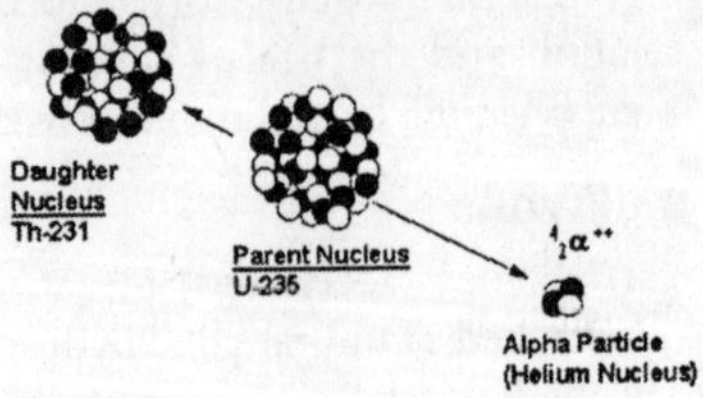

■ alternating copolymer

a copolymer consisting of macromolecules comprising two species of monomeric units in alternating sequence. *Note* An alternating copolymer may be considered as a homopolymer derived from an implicit or hypothetical monomer.

■ alternating copolymerisation

a copolymerisation in which an alternating copolymer is formed.

■ alum

an alum is a double salt comprising a sulphate salt of a monovalent element, with the sulphate salt of a trivalent element. The most well known example is *potash alum* (aluminium potassium sulphate) $\mathbf{KAl(SO_4)_2.12H_2O}$ which is used to clear murky water.

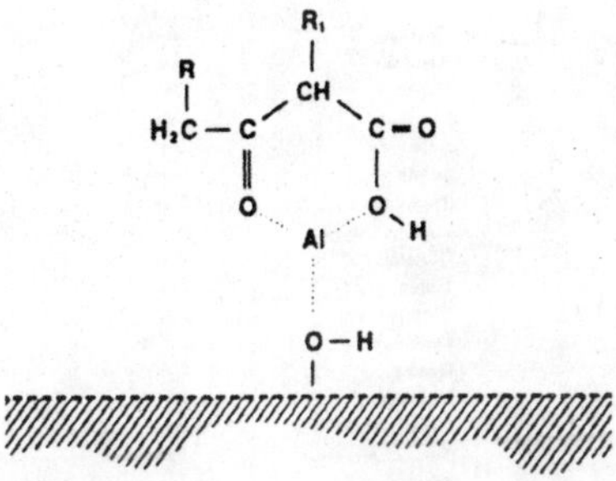

■ alumina

an adsorbent sometimes used in adsorption chromatography. Aluminum oxide (Al_2O_3) is a porous adsorbent that is available with a slightly basic surface. For this reason, it can have advantages over silica, which is considered to have an acidic surface.

■ aluminium

a metallic element used extensively in the manufacture of aircraft components, prosthetic devices and as an ingredient in antiperspirants, antacids and antiseptics.

■ aluminosilicates

silicates containing aluminium, which are a major compound found in many clays, such as kaolin or bentonite.

■ alunite

naturally occurring crystals - only impacts on germs, does not choke pores.

■ amalgam

liquid *mercury* has the ability to dissolve other metals to produce alloys and these alloys are called amalgams. Amalgam examples are *gold/mercury*, *silver/mercury* and *copper/mercury* all of which are used in dentistry.

■ ambergris

a waxy substance, secreted by sperm whales. It is used as a fixative in perfumery.

■ ambient

undisturbed natural conditions such as ambient temperature caused by climate or natural subsurface thermal gradients.

■ Ames test

the Ames Test is used to assess whether a chemical might be a carcinogen. It assumes that carcinogens possess mutagenic activity, and uses bacteria and mammalian microsomes to determine whether a chemical is a mutagen. Approximately 85% of known carcinogens are mutagens. The Ames test, therefore, is a helpful but not perfect predictor of carcinogenic potential.

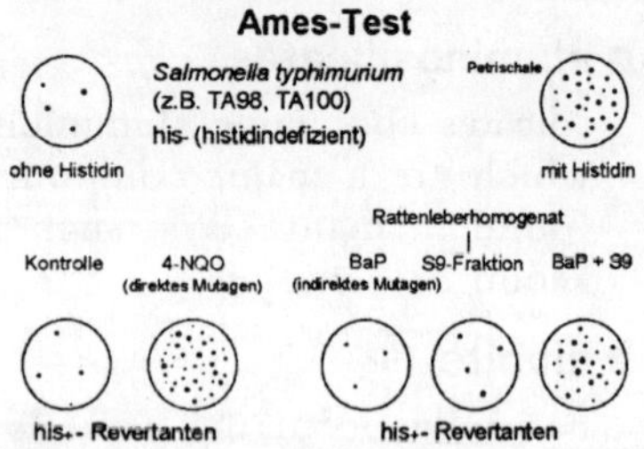

■ amide

1. an organic compound containing the $-CONH_2-$ group (see below). Also see **peptide**.

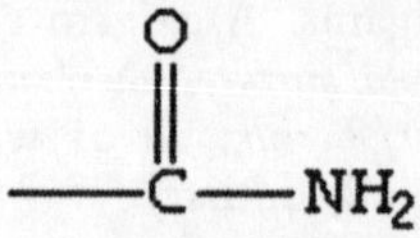

2. an amine is a distinctive smelling organic compound with one or more of the hydrogen atoms in ammonia replaced by organic groups. There are three classes of amines, dependant on the number of hydrogens replaced (see below):
3. the organic groups can be aliphatic, alicyclic or aromatic. Some common amines are listed below:

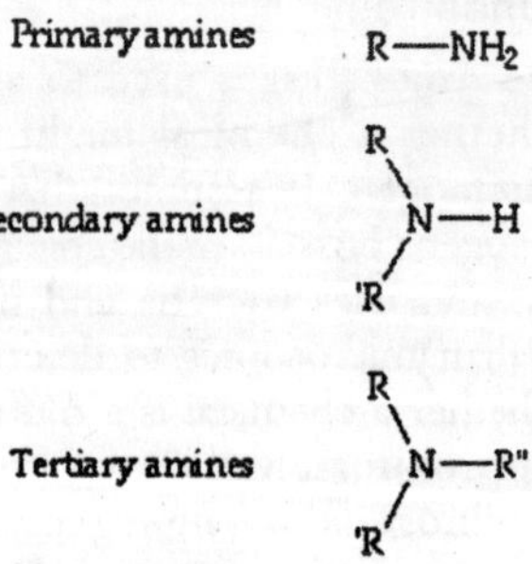

■ amino acid

a large class of substances with molecules including the amino and carboxyl groups. In the human body, amino acids are joined together to form long chains as part of the structures of proteins. Because they possess two functional groups, amino acids exhibit both acidic and basic characteristics. Each amino acid has an abbreviation when in a protein chain, so that the protein **GlyAlaVal** would indicate that a molecule of *Glycine* is joined to one of *Alanine* which is in turn joined to one of *Valine.* A list of the more common amino acids are given below:

The Major Amino acids

Glycine	Gly
Alanine	Ala
Valine	Val
Leucine	Leu
Isoleucine	Iso
Serine	Ser
Threonine	Thr
Aspartic acid	Asp
Glutamic acid	Glu
Lysine	Lys
Arginine	Arg
Cysteine	CySH
Cystine	CySSCy
Methionine	Met
Phenylalanine	Phe
Tyrosine	Tyr
Tryptophan	Try
Histidine	His
Proline	Pro
Hydroxyproline	Hypro

■ amino group

an organic compound containing the $-NH_2-$ group. Substances which contain this group are called amines. Examples include urea, and trimethylamine.

■ amino phase

a propylamino stationary phase used mostly in normal bonded-phase chromatography. It is somewhat reactive for any solute molecule or mobile phase additive that can react with amines. The amino phase has found some applications as a weak anion exchanger.

■ **ammonia**

an alkaline gas composed of nitrogen and hydrogen. Aqueous solutions with 5-10% ammonia are sold as household ammonia.

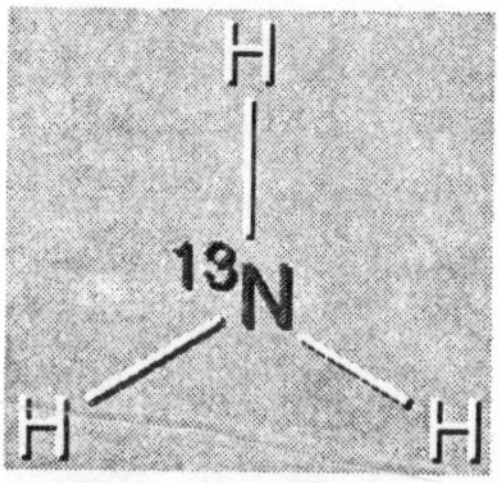

■ **ammonium laureth sulphate**

a synthetic detergent that is used extensively as the primary surfactant in shampoos as a foaming and cleansing agent. Some problems can arise from the ethoxylated detergents.

■ **ammonium lauryl sulphate**

the ammonium salt of sulphated lauryl alcohol, which is used as a foaming and cleansing agent. The absence of any ethoxylation makes it slightly less mild than ammonium laureth sulphate but excludes the problems associated with ethoxylation.

■ **amoebicidal**

for amoebic dysentery.

■ **amorphography**

the branch of science concerned with the determination of amorphous solid structures and their systemmatic classification.

■ **amorphous**

having no long-range order.

■ **amphoteric ion-exchange resin**

ion-exchange resins that have both positive and negative ionic groups. Are most useful for ion retardation in which all ionic materials can be removed from solution; the anionic and cationic functionalities coexist on the same material .

■ **amphoteric surfactant**

a surfactant that, in water solution, may be either anionic or cationic, depending upon the pH.

■ **amygdalin**

a glucoside found in the kernels of most fruit belonging to the Rosaceae family, particularly bitter almondo. It is used widely as a flavouring material in the form of essence of bitter almonds. Although harmless in small doses in its natural form, it becomes toxic when distilled, hence the toxicity of bitter almond oil.

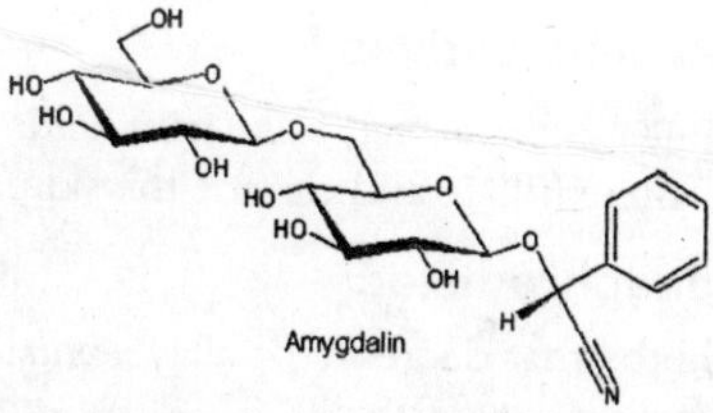

Amygdalin

■ **analgesic**

an Analgesic such as aspirin is a chemical which reduces the body's sensitivity to pain.

analgesic or anodynes

these herbs reduce or eliminate pain (e.g., digestive, circulatory, respiratory, nervous system, nerve, muscle, tooth pain, nervous digestion, headaches). Some herbs are strong pain relievers, often working best against pains of specific causes. Examples: Camphor, chamomile, cinnamon, cloves, echinacea, lavender flower, feverfew herb, cabbage leaves, wintergreen leaf , passionflower herb and flower.

analogue

an analog is a chemical compound that is structurally like another compound except for one element which has been replaced with another element of the same valence from the same group of the periodic table

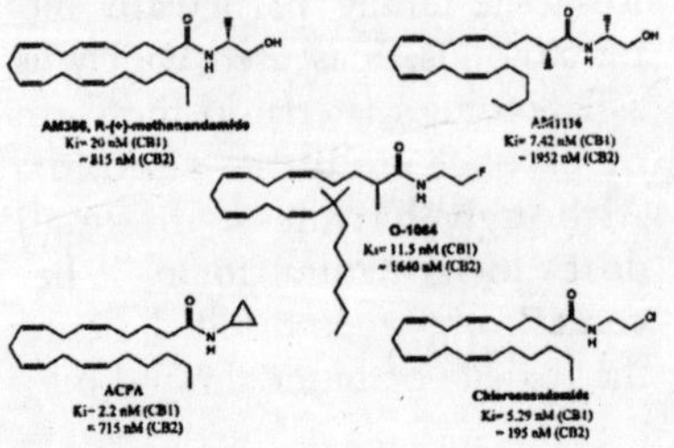

ananas sativus

pineapple extract, mild fruit acids help exfoliate and cleanse the skin.

anaphrodisiac

herbs that decrease or allay sexual feelings or desires.

anesthetics

for surgical anesthesia. Examples are: ashok, calamus, gudmar, and jatamanshi.

angstrom

1. Ångström (Å) is a unit of length equal to 10^{-10} m. The Ångström is defined in terms of the wavelength of the emission spectra's red line of an atom of cadmium (6438.4696 Å).

2. unit definition used to define light wavelength, ultra-violet energy and x-rays; one angstrom is equivalent to 10^{-1} nanometers(1.0 $\times$ 10^{-10} meters)

anhydride

these are compounds formed by the removal of water or *Hydrogen* and *Oxygen* together from another substance. In inorganic chemistry most anhydrides are formed by the loss of the **water of crystallisation**. eg. Copper(II) sulphate is a blue hydrated salt with the formula $CuSO_4.5H_2O$, but if heated, the water is driven off leaving the colourless anhydrous salt $CuSO_4$. In organic chemistry an anhydride is usually the result of the loss of a

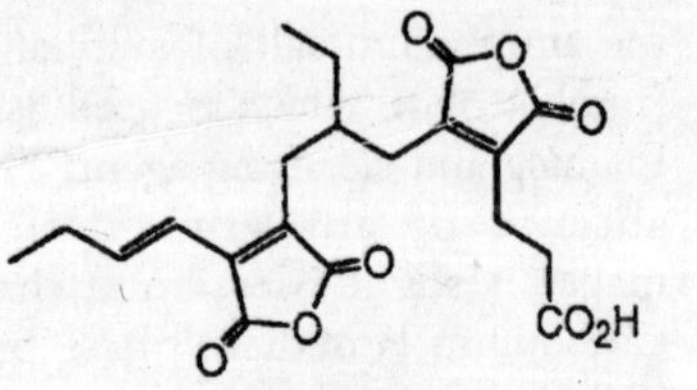

1 (FR222398)

water molecule from a dicarboxylic acid, e.g. from phthalic acid to give phthalic anhydride.

anhydrous

a product that has had all of the water removed.

aniline point

the aniline point of a petroleum product is the minimum equilibrium solution temperature with an equal volume of freshly distilled aniline.

anion

an atom or group of atoms having a negative charge; a negatively charged ion.

Binding Principle of QIAGEN Resin

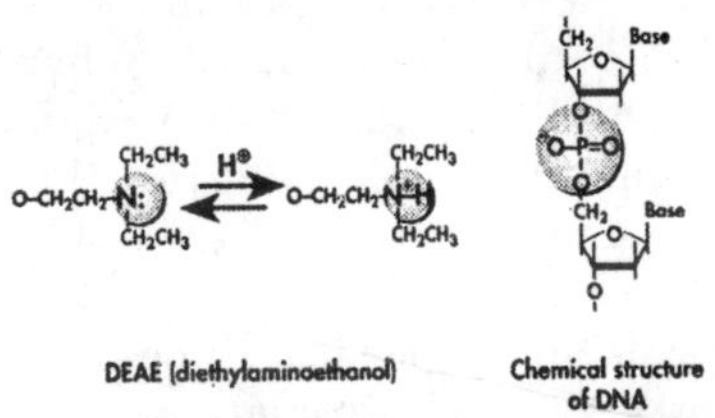

anion exchange

an anionic resin has negative ions built into its structure and therefore exchanges positive ions. In anion exchange, the side groups are ionised basic groups, such as ($-NH_2$, $-NRH$, $-NR_2$, $-NR_3^+$) to which anions OH^- are attached. The exchange reaction is one in which different anions in the solution displace the OH^- from the solid.

anion-exchange chromatography

the ion-exchange procedure used for the separation of anions. Both resins and bonded phases are available for this mode. The tetralkylammonium group is a typical strong anion-exchange functional group. An amino group on a bonded or coated stationary phase would be an example of a weak anion exchanger.

anionic polymerisation

an ionic polymerisation in which the kinetic-chain carriers are anions.

anionic surfactant

negatively charged part of a molecule. Anionic surfactants are widely used in high-sudsing detergents.

anisotropic

exhibiting different values of a property in different crystallographic directions.

anisotropy

the condition in which physical properties vary when measured in different directions or along different axes. For example, in a layered rock section the permeability is often anisotropic in the vertical direction (from layer to layer) but is isotropic in the horizontal direction (within a layer).

annealing

a generic term used to denote a heat treatment wherein the microstructrure and, consequently, the properties of a material are altered. Frequently, refers to heat treatment whereby a cold-worked metal is softened by allowing it to recrystallise.

annual committed effective dose equivalent

a radiation protection term for the effective dose equivalent received by an individual in one year from radiation sources external to the individual plus committed effective dose equivalent.

■ **antiferromagnetism**

a phenomenon observed in some materials in which complete magnetic moment cancellation occurs as a result of antiparallel coupling of adjacent atoms or ions. The macroscopic solid possesses no net magnetic moment.

Antiferromagnetisn

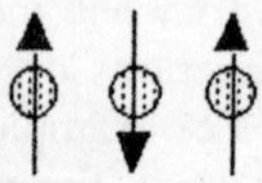

■ **antiparticle**

antiparticle is a particle having the same mass as a given elementary particle and a charge equal in magnitude but opposite in sign.

■ **antiredeposition agent**

an ingredient used in detergents to help prevent soil from redepositing on surfaces or fabrics. Sodium carboxymethylcellulose (CMC) is the most widely used.

■ **antireflection coating**

surface treatment with dielectric material to reduce the reflection of electromagnetic radiation while increasing light transmission.

■ **anti-reset windup**

same as reset windup.

■ **apparent activation energy**

in practice, reaction rates are often determined by physical processes (e.g. mass flow, diffusion, mass transfer area) as well as by chemical processes. The activation energy observed in these cases is called the apparent activation energy.

■ **aqueous**

the literal meaning of the word is 'containing water.' When used in connection with emulsions, it usually indicates that the emulsion is more than 50% water and that other components are suspended in the form of very small droplets, usually surrounded by a surfactant (emulsifying agent), and distributed throughout the water.

■ **aquifer**

a subsurface formation or group of formations that is saturated and sufficiently permeable to transmit quantities of water to wells or springs.

■ **AR**

AR is symbol for element argon.

■ **areal mass loading**

used in thermal loading calculations, the amount of heavy metal (usually expressed in metric tons of uranium or equivalent) emplaced per unit area in the repository.

■ **arenes**

arenes are monocyclic and polycyclic aromatic hydrocarbons.

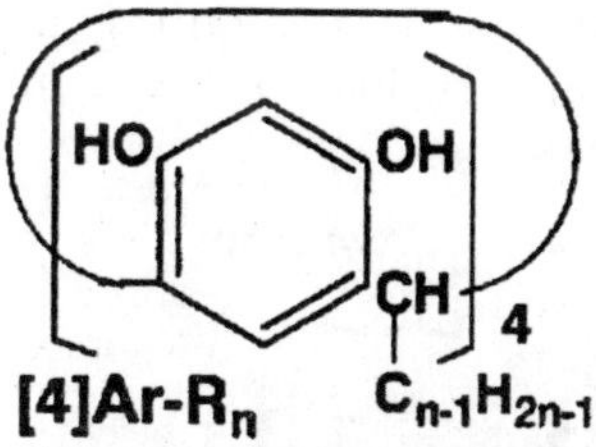

R_n: C_4, C_5, C_6, C_7, C_8, C_{10}, C_{12}, C_{16}, C_{18}

■ aromatic compounds

1. aromatic compounds are major group of unsaturated cyclic hydrocarbons containing one or more rings, typified by benzene, which has a 6-carbon ring containing three double bonds. All the bonds in benzene (C_6H_6) are the same length intermediate between double and single C-C bonds. The properties arise because the electrons in the p-orbitals are delocalised over the ring, giving extra stabilisation energy of 150 kJ/mol over the energy of Kekulé structure. Aromatic compounds are unsaturated compounds, yet they do not easily partake in addition reactions.
Historical use of the term implies a ring containing only carbon (e.g., benzene, naphthalene), but it is often generalised to include heterocyclic structures such as pyridine and thiophene.
2. group of hydrocarbons of which benzene is the parent. They are called 'aromatics' because many of their derivatives have sweet or aromatic odors.

■ Arrhenius plot

graph that shows the logarithmic rate of heat production ln(q) versus the inverse temperature (1/T) in Kelvin. This results in a straight line with a gradient -Ea/R. In practice, the rate of reaction is often affected by physical processes (e.g. diffusion or mass transfer), which results in the occurrence of an apparent activation energy.

■ Arrhenius rate equation

the rate coefficient for a chemical reaction (k) increases exponentially with increasing temperature (T). This relationship is represented by the Arrhenius equation k = A exp (-Ea/RT).

Balanced Equation:

$$2\,IO_3^- + 5\,SO_3^- + 2\,H^+ \longrightarrow I_2 + 5\,SO_4^{2-} + H_2O$$

Oxygen atom count		Charge count	
LHS	RHS	LHS	RHS
2x3 5x3	5x4 1	-2-10+2	-10
21	21	-10	-10

■ ash

fine or very fine pyroclastic particles, less than 4 mm (0.15 in.) in diameter, that are blown out from a volcanic explosion.

■ ash content

non-combustible residue of a lubricating oil (also fuels) determined in accordance with ASTM D582 - also D874 (sulphated ash). Since some detergents are metallic salts or compounds, the percentage of ash has been considered to have a relationship to detergency. Interpretations can be grossly distorted, however, for the following reasons: 1. Detergency depends on the properties of the base oil as well as on the additive. Some combinations of base oil and additive are much more effective than others. 2. Detergents vary considerably in their potency, and some leave more ash than others. Organic detergents have been developed, in fact, that leave no ash at all. 3. Some of the ash may be contributed by additives other than detergents. 4. There appears to be a limit to the effective concentration of detergent. Nothing is gained by exceeding this limit, and a superabundance of detergent may actually reduce cleanliness.

■ astronomical unit

astronomical unit (AU) is a unit of length employed in astronomy for describing planetary distance. It is the mean distance of the earth from the sun, equal to $1.49597870 \times 10^{11}$ m.

■ asymmetry

factor describing the shape of a chromatographic peak. Theory assumes a Gaussian shape and that peaks are symmetrical. The peak asymmetry factor is the ratio (at 10% of the peak height) of the distance between the peak apex and the back side of the chromatographic curve to the distance between the peak apex and the front side of the chromatographic curve. A value > I is a tailing peak, while a value <I is a fronting peak.

Asymmetry

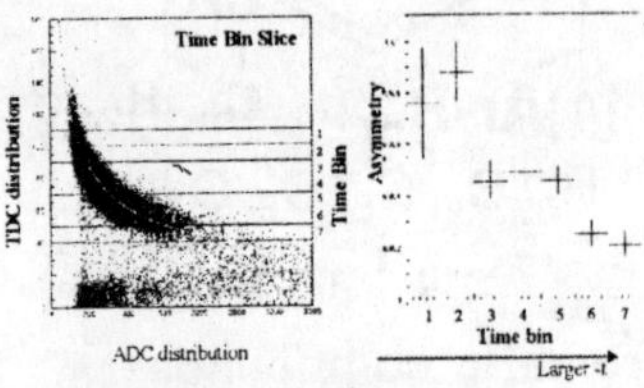

Asymmetry decreases with t value.
Consistent with theory prediction qualitatively.

■ atactic

a type of polymer chain configuration wherein side groups are randomly poitioned on one side of the polymer backbone or the other.

■ atactic macromolecule

a regular macromolecule in which the configurational (base) units are not all identical.

■ atactic polymer

a polymer composed of atactic macromolecules .

■ a-term

the first term in the van Deemter equation.

■ atom

the smallest particle of an element that retains the chemical properties of that element. The atoms of many elements are bonded together in groups to form particles called molecules. Atoms consist of three main types of smaller particles. These include the electrons, protons and neutrons.

■ **atomic number**

atomic number (Z) is a characteristic property of an element, equal to the number of protons in the nucleus.

■ **atomisation characteristics**

the ability of an oil to be broken up into a fine spray by some mechanical means.

■ **austenite**

face-centred cubic iron; also iron and steel alloys that have the FCC structure.

■ **auto mode**

in auto mode the controller calculates the output using the error signal (difference between setpoint and PV).

■ **autocatalysis**

the increase of the rate of reaction due to the catalysing effect of the reaction products.

■ **autocatalytic criticality**

a transient criticality in which the usual mechanisms that tend to shut down a criticality are delayed until a high fission rate is achieved.

■ **autoignition temperature**

autoignition temperature is the minimum temperature required to initiate or cause self-sustained combustion in any substance in the absence of a spark or flame. This varies with the test method.

■ **Avogadro constant**

Avogadro constant (N_A or L) is the number of elementary entities in one mole of a substance.

$L = N / n$

It has the value (6.022 045 ± 0.000 031) × 10^{23} mol^{-1}.

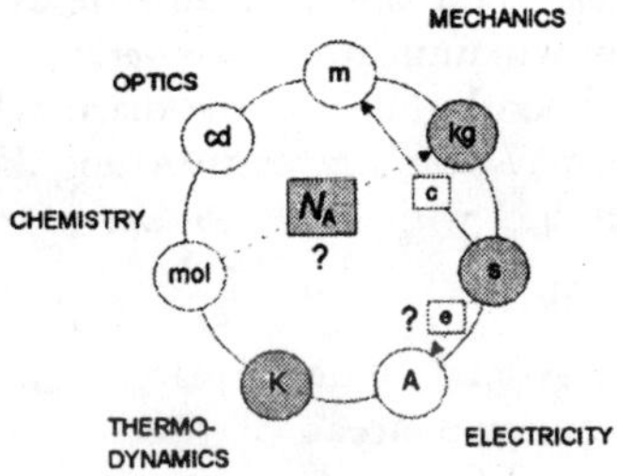

■ **azeotrope**

azeotrope is a mixture of two liquids that boils at constant composition, i.e. the composition of the vapour is the same as that of the liquid. Azeotropes occur because of deviations in Raoult's law leading to a maximum or minimum in the boiling point - composition diagram. The composition of an azeotrope depends on the pressure.

■ **azeotropic drying**

method of removing water from a liquid at temperatures lower than 100°C; a second liquid that forms an azeotropic mixture with water is added to the sample liquid.

■ **azo compounds**

azo compounds are organic compounds containing the group -N=N- linking two other groups. They can be formed by reaction of a diazonium ion with a benzene ring.

■ **azo dye**

an azo dye is one of an extensive range of synthetic organic dye made from *aniline*, by first converting it with *sodium nitrite* to a *diazonium chloride salt* which is then reacted with other *aromatic amines*, *phenols* and *sulphonic acids*. The well known indicator *methyl-orange* is formed by making the diazo salt of *sulphanilic acid*, and reacting the product with *dimethylaniline*.

■ **backfill**

the general fill that is placed in the excavated areas of the underground facility. If used, the backfill for the repository may be tuff or other material.

■ **backflushing**

a column-switching technique in which a four-way valve placed between the injector and the column allows for mobile phase flow in either direction. Backflushing is used to elute stronglyheld compounds at the head of the column. It can be used to analyse these compounds or merely to remove them from the column.

■ **background radiation**

radiation arising from natural radioactive material always present in the environment, including solar and cosmic radiation, radon gas, soil and rocks, food, and the human body.

■ **background**

the surface area against which the pattern is relieved. The surface area of the background is usually much greater than the area relieved.

■ **backlighting**

viewing or photographing an object by placing it between a light-source & the eye or recording medium.

■ **bainite**

an Fe-C composition consisting of a fine dispersion of cementite in alpha-ferrite. It is an austenitic transformation product that forms at temperatures between those at which pearlite and martensite transformations occur.

■ **bake**

a thermal process used to dry or cure.

■ **ball mills**

ball mills, also known as centrifugal or planetary mills, are devices used to rapidly grind materials to colloidal fineness (approximately 1 micron and below) by developing high grinding energy via centrifugal and/or planetary action.

■ **band broadening**

the dilution of the chromatographic band as it moves down the column. The peak is injected as a slug, and, if not for the process of band broadening, each separated component would elute as a narrow slug of pure compound. The measure of band broadening is band width, t_w, or more correctly, the number of theoretical plates in the column. N.

■ **band dispersion**

see **band broadening**

■ **band gap**

the energy difference between the top of one band and the bottom of the next higher band.

■ **band gap energy**

for semiconductors and insulators, the energies that lie between the valence and conduction bands.

■ **band spreading**

see **band broadening**

■ **band width (t_w)**

the width of the chromatographic band during elution from the column. It is usually measured at the baseline by drawing tangents to the sides of the Gaussian curve representing the peak. Small band widths usually represent efficient separations. Also referred to as peak width.

■ **bar**

bar (bar) is a unit of pressure equal to 10^5 Pa. Its use is temporarily maintained with the SI. The milibar (100 Pa) is commonly used in meteorology.

■ **barotropic**

a barotropic fluid is defined as that state of a fluid for which **ñ** is a function of only the pressure. The condition of barotropy of a fluid represents another rather idealised state. However, in this case the situation is closer to reality since compressibility is allowed for.
The term 'barotropic' infers 'turning with (or in the same manner as) the isobars', referring to the isopycnals. The name is a lucid one since it is obvious that if ñ depends only on **p** then the isopycnal surfaces must always be parallel to the isobaric surfaces, hence any change in inclination of the latter brings about an identical change in orientation of the isopycnal surfaces.

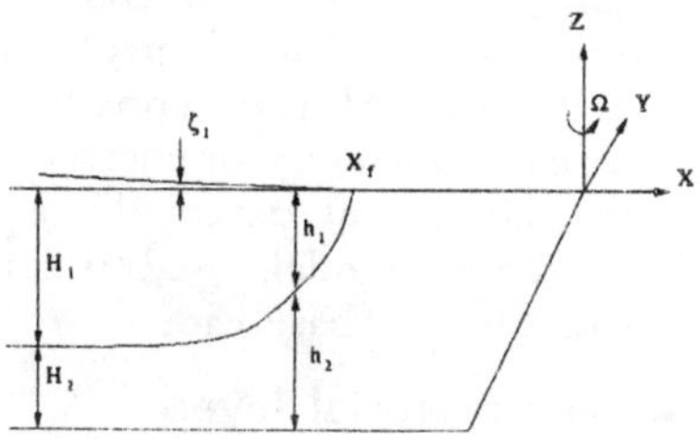

■ **barrier**

any material, structure, or feature that prevents or substantially reduces the rate of movement of water or radionuclides.

■ **baryon**

a particle made of three quarks (Protons and Neutrons).

■ **base**

historically, base is a substance that yields an OH - ion when it dissociates in solution, resulting in a pH>7. In the Brønsted definition, a base is a substance capable of accepting a proton in any type of reaction. The more general definition, due to G.N. Lewis, classifies any chemical species capable of donating an electron pair as a base. Typically, bases are metal oxides, hydroxides, or compounds (such as ammonia) that give hydroxide ions in aqueous solution.

■ **base case**

the sequence of anticipated conditions expected to occur in and around the potential repository, without the inclusion of unlikely or unanticipated features, events, or processes. The components that contribute to the base case model are intended to encompass this probable behaviour of the repository, based on the range of uncertainty for the various parameters and conceptual models used in constructing the base case.

■ **base material layer**

the material layer onto which coating or plating is applied, and or from which stock is selectively removed by photo chemical machining; onto which a strippable photosense is applied for use with phototools.

■ **base pairs**

the parts that make up the rungs in the ladder structure of the DNA double helix.

■ **batch process**

an operation in which all ingredients are added to a reactor at the start of a reaction.

■ **Becquerel**

Becquerel (Bq) is the SI derived unit, with a special name, for radioactivity, equal to s^{-1}. It describes a radioactivity of an amount of radionuclide decaying at the rate, on average, of one spontaneous nuclear transition per second. The unit is named after the French scientist A. H. Becquerel (1852-1908).

■ **Beer's law**

Beer's law is the functional relationship between the quantity measured in an absorption method (A) and the quantity sought, the analyte concentration (c). As a consequence of interactions between the photons and absorbing particles, the power of the beam is attenuated from P_0 to P. Beer's law can be written

$A = \log (P_0/P) = abc$

where a is a proportionality constant called the absorptivity and b is the path length of the radiation through the absorbing medium. Also called the Beer-Lambert law.

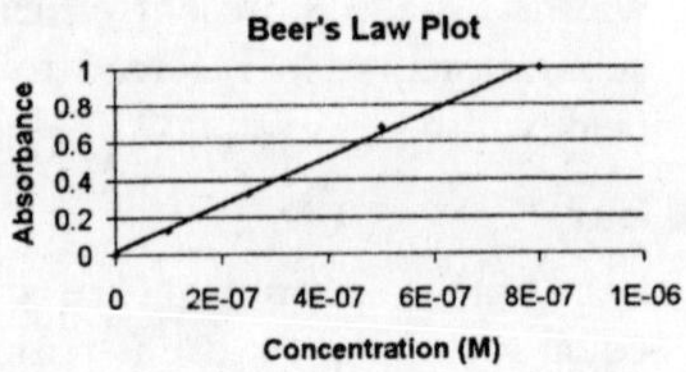

■ **bentonite**

a form of clay that swells in the presence of water to give a gel. It is used as a thickening agent and is considered non-toxic, however, some people claim that it may clog the pores and suffocate the skin.

■ **benzene**

a petrochemical used as a solvent and manufacturing agent in lacquers, varnishes and cosmetics. Highly toxic in even minute proportions, it has known carcinogenic activity.

■ **benzoic acid**

a white powder that is used as an antimicrobial agent both in food products and in toiletries. It is slightly toxic and can cause skin irritations.

■ **BET method**

a method developed by Bruner, Emmett, and Teller for measuring surface area by using nitrogen adsorption condensation in pores at liquid nitrogen temperature. Pore volume and pore-size distribution can also be obtained by this method.

■ **beta particles**

beta particle is a charged particle emitted from a radioactive atomic nucleus either natural or manufactured. The energies of beta particles range from 0 to 4 MeV. They carry a single charge; if this is negative, the particle is identical with an electron; if positive, it is a positron. An unstable atomic nucleus changes into a nucleus of the same mass number but different proton number with the emission of an electron and an antineutrino (or a positron and a neutrino)

■ **beta radiation**

streams of beta particles are known as beta ray or beta radiation. Beta rays may cause skin burns and are harmful within the body. A thin sheet of metal can afford protection to the skin.

■ **bifunctional monomer**

a monomer unit that has two active bonding positions.

■ **biochip technologies**

multiple element DNA array devices directed at gene expression, genotyping, pharmacogenetics, and infectious disease diagnostics. Multiple element array devices using any form of biomolecular recognition or interaction and formed using techniques developed and used in microelectronics manufacture. Integrated, micro-scale lab-on-a-chip instruments for research, drug discovery and point-of-concern analysis. Integrated, micro-scale devices for the controlled release of bioactive molecules (drugs, hormones, growth factors, etc.)

■ **biochips-type i**

devices that are formed using microelectronics fabrication techniques and technologies and are given to a biological function. Example, a sub-coetaneous transmitter-receiver for tracking an animal using the GPS.

■ **biochips-type ii**

high density arrays of DNA and RNA-probes for the detection and measurement of DNA and RNA sequences. These are effectively multi-element array biosensors that are engineered using microelectronics fabrication techniques and technologies. Example, the Affymetrix P53 gene chip.

■ **biochips-type iii**

are bio-molecular switching devices that use the natural order and templating characteristics of biologically active molecules such as DNA to build complex structures capable of transistor-like switching function. These are molecular electronic devices or molecular computers.

■ **biocompatible**

term to indicate that the column or instrument component will not irreversibly or strongly adsorb or deactivate biomolecules such as proteins.

■ **bioelectronics**

a fundamental sub-discipline that embraces the study of electron and ion transport phenomena, signaling and signal transduction, and the associated molecular and organisational structures that control and influence these in living systems. Bioelectronics may be studied at the whole animal, organ, tissue, cellular, and sub-cellular levels.

■ **bioinformatics**

the development and application of software tools to the analysis, data mining, and interpretation of large databases (e.g. those developed from genomics research and drug screenings), to derive and present useful information.

■ **bionic**

having anatomical structures or physiological processes that are replaced or enhanced by electronic or mechanical components.

■ **biosensor technologies**

affinity binding for ligand fishing and biomolecular interaction studies. Point of concern diagnostics and monitoring.

■ **biosensors**

artificial test, measurement, and monitoring devices that use bioactive indicator molecules, organelles, cells and tissues as the recognition component in conjunction with solid state physicochemical transducers. These are analytical devices. Biosensors may be single-element or multi-element, may target a single analyte or multiple analytes. Often the term biosensor is used to describe only the detector.

■ **biosphere**

the ecosystem of the earth and the living organisms inhabiting it.

■ **biosphere dose conversion factor**

a multiplier used in converting a radionuclide concentration at the geosphere/biosphere interface into a dose that a human would experience for all pathways considered, with units expressed in terms of annual dose (i.e., the total effective dose equivalent) per unit concentration.

■ **black body radiation**

black body radiation is the radiation emitted by a perfect black body, i.e., a body that absorbs all radiation incident on it and reflects none. The wavelength dependence of the radiated energy density *ñ* (energy per unit volume per unit wavelength range) is given by the Planck formula

■ **blackbody**

in radiation physics, an ideal blackbody is a theoretical object that absorbs all the radiant energy falling upon it and emits it in the form of thermal radiation. Planck's radiation law gives the power radiated by a unit area of blackbody, and the Stefan-Boltzman law expresses the total power radiated.

■ **block**

a portion of a macromolecule, comprising many constitutional units, that has at least one feature which is not present in the adjacent portions.

■ **block copolymer**

a copolymer that is a block polymer . In the constituent macromolecules of a block copolymer, adjacent blocks are constitutionally different, i.e., adjacent blocks comprise constitutional units derived from different species of monomer or from the same species of monomer but with a different composition or sequence distribution of constitutional units.

■ **block macromolecule**

a macromolecule which is composed of blocks in linear sequence.

■ **block polymer**

a polymer composed of block macromolecules.

■ **block-bounding fault**

a normal fault that divides the crust into structural or fault blocks of different elevations and orientations; typical of the Basin and Range province.

■ **blowdown**

rapid discharge of a vessel

■ **body-centreed cubic (bcc)**

a common crystal structure that contains atoms located at the corners of a cubic cell and one atom at the cell centre position.

■ **Bohr magneton**

Bohr magneton ($\hat{\imath}_B$) is the atomic unit of magnetic moment, defined as

$\hat{\imath}_B = eh/4\eth m_e = 9.274\times10^{-24}$ Am^2

where *h* is Planck's constant, m_e the electron mass, and *e* the elementary charge. It is the moment associated with a single electron spin.

■ **boiling point**

boiling point is the temperature at which the liquid and gas phases of a substance are in equilibrium at a specified pressure. The normal boiling point is the boiling point at normal atmospheric pressure (101.325 kPa).

■ **boiling-water reactor**

a nuclear power reactor in which water passing as coolant through the core is turned to steam by direct use of fission heat from the uranium oxide fuel; steam for driving the turbogenerator is formed within the reactor vessel itself rather than in an external heat exchanger and, after being condensed, returns as feedwater to the reactor vessel.

■ Boltzmann constant

Boltzmann constant (k) is the molar gas constant *R* divided by Avogadro's constant. It has the value 1.380658×10^{-23} J/K.

■ bomb-pulse

a detectable increase in radionuclide concentration related to global fallout from above-ground thermonuclear tests.

■ bonded-phase chromatography (bpc)

the most popular LC mode. A stationary phase chemically bonded to a support is used for the separation. The most popular support is microparticulate silica gel. and the most popular type of bonded phase is the ganosilane, such as octadecyl (for reversed-phase chromatography). Approximately 70% of all HPLC is carried out on chemically bonded phases.

■ bonding energy

the energy required to separate two atoms that are chemically bonded to each other.

■ borehole

a hole drilled for purposes of collecting site characterisation data or for supplying water. Sometimes referred to as a drillhole or well bore.

■ Born-Haber cycle

Born-Haber cycle is a cycle of reactions used for calculating the lattice energies of ionic crystalline solids. For a compound MX, the lattice energy is the enthalpy of the reaction

$M^+(g) + X^-(g) \rightarrow M^+X^-(s)$ ÄH$_L$

The standard enthalpy of formation of the ionic solid is the enthalpy of the reaction

$M(s) + 1/2X_2(g) \rightarrow M^+X^-(s)$ ÄH$_f$

The cycle involves equating this enthalpy (which can be measured) to the sum of the enthalpies of a number of steps proceeding from the elements to the ionic solid.

■ borosilicate glass

a material used to vitrify high-level radioactive waste in which boron takes the place of the lime used in ordinary glass mixtures.

■ boundary condition

for a model, the establishment of a set condition (set value), often at the geometric edge of the model, for a given variable; for example, using a specified groundwater flux from infiltration as a boundary condition for a flow model.

■ boundary lubrication

a state of lubrication characterised by partial contact between two metal surfaces, and partial separation of the surfaces by a fluid film of lubricant.

■ bounding

for a mathematical model, relating to a set condition (set value) that is considered to reflect the reasonable extreme for that value in the real-world condition being modelled.

■ **bounding value**

specific data point that defines the reasonable extreme limit of a variable in an experiment or model.

■ **bourdan gauge**

a device used to measure the pressure in which the pressure to be measured is applied to a curved tube, oval in cross section. Pressure applied to the tube tends to cause the tube to straighten out, and the deflection of the end of the tube is communicated through a system of levers to a recording needle. This gauge is widely used for steam and compressed gases. The pressure indicated is the difference between that communicated by the system to the external (ambient) pressure, and is usually referred to as the gauge pressure.

■ **Boyle's law**

Boyle's law is the empirical law, exact only for an ideal gas, which states that the volume of a gas is inversely proportional to its pressure at constant temperature.

■ **Bragg angle**

Bragg angle (è) is defined by the equation

në = 2dsinè

which relates the angle è between a crystal plane and the diffracted x-ray beam, the wavelength *ë* of the x-rays, the crystal plane spacing *d*, and the diffraction order *n* (any integer).

■ **Bragg's law**

a relationship that stipulates the condition for diffraction by a set of crystallographic planes.

■ **branch**

side-chain pendant chain An oligomeric or polymeric offshoot from a macromolecular chain . An oligomeric branch may be termed a short-chain branch. A polymeric branch may be termed a long-chain branch.

■ **branch point**

a point on a chain at which a branch is attached.

■ **branch unit**

a constitutional unit containing a branch point .

■ **branched chain**

a chain with at least one branch point intermediate between the boundary units.

■ **branched polymer**

a polymer having a molecular structure of secondary chains that extend from the primary chains.

■ **brass**

a copper-rich copper-zinc alloy.

■ **brazing**

a metal joining technique that uses a molten filler metal alloy having a melting temperature greater than about 425 °C.

■ **breakthrough**

the time at which the concentration of a substance, usually in groundwater, arrives at a particular point of interest.

■ **breakthrough curve**

the curve describing the rate of arrival of radionuclides transmitted through a medium. The breakthrough curve calculation includes the effects of all flow modes, flow in rock matrix, flow in fractures, and retardation and determines the expected proportion of the radionuclide mass transmitted through the medium as a function of time.

■ **breakthrough time**

the breakthrough time is the time taken in standard tests for permeation of a chemical through a protective barrier (such as a rubber glove) to be detected.

■ **breakthrough volume**

the volume at which a particular solute pumped continuously through a column will begin to elute The breakthrough volume is useful in determining the total sample capacity of the column for a particular solute.

■ **bridge crane**

a large overhead crane used for material handling that spans across rails on either side of a structure.

■ **brittle fracture**

fracture that occur by rapid crack propagation and without appreciable macroscopic deformation.

■ **bronze**

a copper-rich copper-tin alloy.

■ **Brookfield viscosity:**

viscosity, in centipoises, as determined on the Brookfield viscometer (ASTM D2983). The operating principle for the Brookfield viscometer is the torque resistance on a spindle rotating in the fluid being tested. Although Brookfield viscosities are most frequently associated with low temperature properties of gear oils and transmission fluids, they are in fact determined for many other types of lubricants.

■ **Brownian motion**

Brownian motion is the continuous random movement of small particles suspended in a fluid, which arise from collisions with the fluid molecules. First observed by the British botanist R. Brown (1773-1858) when studying pollen particles. The effect is also visible in particles of smoke suspended in a still gas.

■ **bs & w monitor**

an instrument which detects entrained water content in petroleum products wherein the water changes the capacitive reactance as a function of the dielectric constant.

■ **bs & w**

bottom sediment and water.

■ **b-term**

same as molecular diffusion term.

■ **bubble point**

upon heating a liquid mixture, this is the point at which bubbles first appear.

■ **buckyball**

a molecule made from carbon arranged into a ball shape. The full name of this molecule is 'buckminsterfullerene'. It is made from 60 carbon atoms bonded together in the same shape as a soccer ball.

buffer

buffer is a solution designed to maintain a constant pH when small amounts of a strong acid or base are added. Buffers usually consist of a fairly weak acid and its salt with a strong base. Suitable concentrations are chosen so that the pH of the solution remains close to the pK_a of the weak acid.

builder

a material that upgrades or protects the cleaning efficiency of a surfactant. Builders inactivate water hardness, supply alkalinity to assist cleaning, provide buffering to maintain alkalinity, prevent redeposition of soil and emulsification of oily and greasy soils.

build-up

a heavy deposit of floor finish, wax, dirt and grime. It is caused by adding layer after layer of floor finish over dirt without deep scrubbing the old layers away first. These build-ups are frequently found along baseboards and corners.

bullseye

stylised control pattern located within the borders areas of a (phototool or photomask or master) to assist in phototooling alignment and/or control registration.

bunker fuel oil

heavy, residual fuel oil used in ships.

burgers vector

a vector that denotes the magnitude and direction of lattice distortion associated with a dislocation.

burn in

to heat a printed image until the film resist coating becomes photo chemical etch resistant or the burn in process utilised to complete fineline image developing of photoresist.

burnup

a measure of nuclear-reactor fuel consumption expressed either as the percentage of fuel atoms that have undergone fission or as the amount of energy produced per initial unit weight of fuel.

butyl paraben

butyl p-hydroxybenzoate.

butyrospermum parkii

shea butter - an emollient which spreads easily, has healing properties and can help protect skin from UV-radiation.

c/h ratio:

carbon/hydrogen ratio

c4, c8, c18

refers to the alkyl chain length of a reversed bonded phase.

CAD

computer aided designing.

calamine

a pink powder made of zinc oxide with a small amount of ferric oxide. It is a traditional mixture used tor relieving itchy skin and rashes. It is not generally considered to be toxic.

■ calcination

a high-temperature reaction whereby one solid material dissociates to form a gas and another solid.

■ calcium carbonate

commonly known as chalk, a naturally occurring calcium salt found in limestone, coral and marble.

■ calcium carbonate

an inorganic compound that occurs naturally as chalk and limestone. Its very slight solubility in water is a chief cause of 'hardness' in water.

■ caldera

a large basin-shaped volcanic depression, more or less circular, resulting from the collapse of the ground surface following a rapid volcanic eruption. A caldera may be more than 15 km (9 mi) in diameter and more than 1,000 m (3,300 ft) deep.

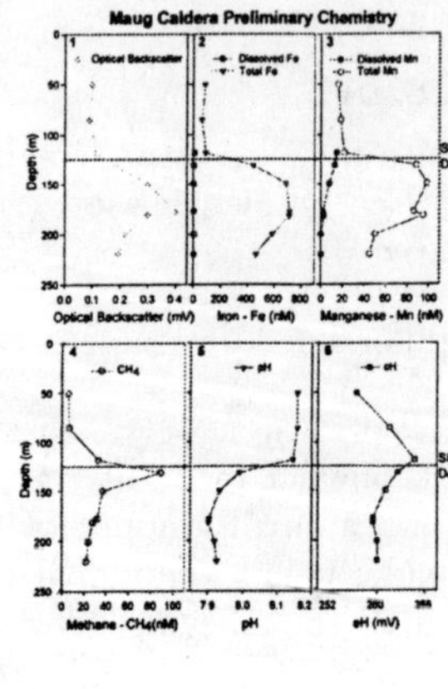

■ calibration

1. in modelling, the process of comparing the conditions, processes, and parameter values used in a model against actual data points or interpolations from measurements at or close to the site to ensure that the model is compatible with 'reality' to the extent feasible.
2. for tools used for field or lab measurements, the process of taking instrument readings on standards known to produce a certain response to check the accuracy and precision of the instrument.

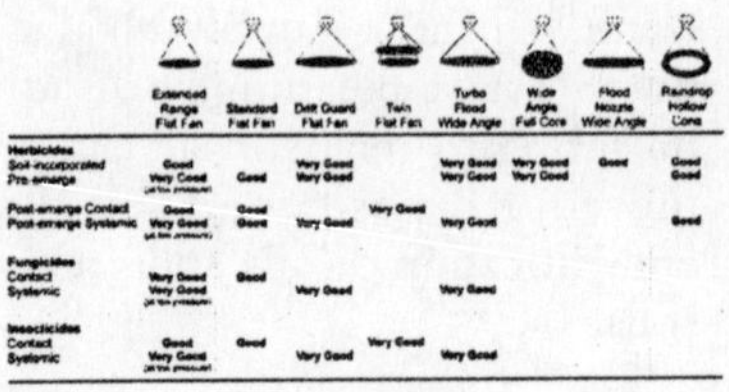

■ calomel electrode

calomel electrode (calomel half cell) is a type of half cell in which the electrode is mercury coated with calomel (Hg_2Cl_2) and the electrolyte is a solution of potassium chloride and saturated calomel. In the calomel half cell the overall reaction is

$Hg_2Cl_2(s) + 2e^- \; 2Hg(l) + 2Cl^-$

■ CAM

computer aided manufacturing.

■ camera reduction

the process involved in photographically reducing the scaled artwork size; the product produced by such a process.

candela

candela (cd) is the SI base unit of luminous intensity.
The candela is the luminous intensity, in a given direction, of a source that emits monochromatic radiation of frequency 540×10^{12} Hz and that has a radiant intensity in that direction of 1/683 watt per steradian.

canine genome project

a project analogous to the human genome project with the goal of mapping the entire genetic code of the canine species

canister

the structure surrounding some forms of waste (e.g., high-level radioactive waste immobilised in glass logs or ceramic disks within cans) that facilitates handling, storage, transportation, and/or disposal.

cannabis saliva

hemp seed oil, naturally rich in minerals and fatty acids which are beneficial for hair and skin.

capacitance

the charge-storage ability of a capacitor, defined as the magnitude of charge stored on either plate divided by the applied voltage.

capacity

see **sample capacity.**

capacity factor

a chromatographic parameter that measures the degree of retention.

capillarity

the general behaviour of fluids acting with surface tension on interfaces or boundaries.

capillary barrier

a contact in the unsaturated zone between a geologic unit containing relatively small-diameter openings and a unit containing relatively large-diameter openings. Water does not flow from the former to the latter due to capillary forces.

capillary drying

the evaporation of moisture from the surface of a porous mass and the subsequent capillary movement of moisture from the interior of the mass to its surface; eventually the amount of moisture on the surface and interior of the mass stabilises.

capillary force

a phenomenon that results from the force of molecular attraction (adhesion) between a fluid and different solid materials; this force causes water to rise in small diameter tubes and, in combination with the effects of gravity, is a means of water movement in the unsaturated zone.

capillary pressure

the pressure due to capillary forces, i.e., the pressure of fluids under the influence of surface tension and adhesion.

capillary tubing

tubing for connecting various parts of the chromatograph. Most capillary tubing used in HPLC is <0.020 in. i d. The smallest useful i d. is about 0.004 in.

■ capping

same as endcapping

■ carbon nanotube

a molecule made from carbon atoms connected into a tube and sealed at both ends by a structure that is like half a buckyball. The tubes may be as long as a few millimeters. Also called 'buckytubes'.

■ carbonyl group

an organic compound containing the >CO group (see below). When a hydrogen atom is attached to the carbon, the resulting compound is known as an aldehyde. When

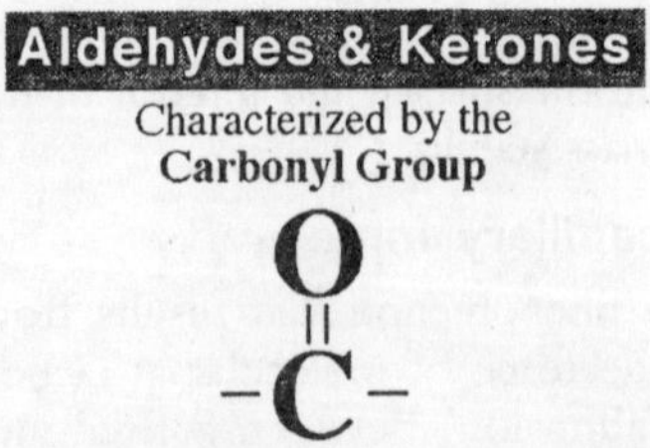

only carbon atoms are attached, the resulting compound is known as a ketone.

■ carboxyl group

an organic compound containing the -COOH group.

■ carboxylic acid

1. a substance containing the carboxyl group. The carboxylic hydrogen can be lost as a hydrogen ion, so these substances are also acidic. An example includes acetic acid.
2. carboxylic acids are organic compounds characterised by the presence of one or more RC(=O)OH groups (the carboxyl group). In systematic chemical nomenclature carboxylic acids names end in the suffix -oic (e.g. ethanoic acids, CH_3COOH). The carbon of the terminal group being counted as part of the chain. They are generally weak acids. Carboxylic acids include the large and important class of fatty acids and may be either saturated or unsaturated. There are also some natural aromatic carboxylic acids (benzoic, salicylic).

■ carburising

the process by which the surface carbon concentration of a ferrous alloy is increased by diffusion from the surrounding environment.

■ carnot cycle

carnot cycle is the most efficient cycle of operations for a reversible heat engine. Published in 1824 by N. L. S. Carnot (1796-1832), it consists of four operations on the working substance in the engine:
a) Isothermal expansion at thermodynamic temperature T_1 with heat q_1 taken in.
b) Adiabatic expansion with a fall of temperature to T_2.
c) Isothermal compression at temperature T_2 with heat q_2 given out.
d) Adiabatic compression at temperature back to T_1.
According to the Carnot principle, the efficiency of any reversible heat engine depends only on the temperature range through

which it works, rather than the properties of the working substances.

■ carrier

a term used most often in affinity chromatography. Refers to the support that is used to carry the active ligand, usually by a covalent bond. Can also refer to the Support in other chromatographic modes .

■ cartridge column

a type of column that has no end fittings and is held in a cartridge holder. The column is a tube; the packing is contained by frits in each end of the tube. Cartridges are easy to change and are less expensive and more convenient than conventional columns with end fittings.

■ cascade

with 2 or more controllers. The output of the 'Master' controller is the setpoint for the 'Slave' controller. A classic example is the control of a reactor (a large vessel with a steel jacket around it). The product temperature (master) controller's output is the setpoint of the jacket temperature (slave) controller.

■ cask

a large, shielded container for shipping or storing spent nuclear fuel and/or high-level radioactive waste that meets all applicable regulatory requirements.

■ cast iron

a ferrous alloy with carbon content between 2 and 4.5 wt%.

■ cat cracker

a large refinery vessel for processing reduced crudes or other feedstocks in the presence of a catalyst, as opposed to the older method of thermal cracking, which employs heat and pressure only. Catalytic cracking is generally preferred since it produces less gas and other highly volatile byproducts. It produces a motor fuel of higher octane than the thermal process.

■ catalyst

catalyst is a substance that increases the rate of a chemical reaction without itself undergoing any permanent chemical change. Catalysts that have the same phase as the reactants are homogeneous catalysts (e.g. enzymes in biochemical reactions). Those that have a different phase are heterogeneous catalysts (e.g. metals or oxides used in gas reactions).

The catalyst provides an alternative pathway by which the reaction can proceed, in which the activation energy is lower. It thus increases the rate at which the reaction comes to equilibrium, although it does not alter the position of the equilibrium.

■ cathode

cathode is a negative electrode of an electrolytic cell to which positively charged ions (cations) migrate when a current is passed as in electroplating baths. In a primary or secondary cell (battery or accumulator) the cathode is the electrode that spontaneously becomes negative during discharge, and from which therefore electrons emerge.

In vacuum electronic devices, electrons are emitted by the cathode and flow to the anode.

■ cathodic protection

a means of corrosion prevention whereby electrons are supplied to the structure to be protected from an external source such as another more reactive metal or a dc power supply.

■ cation

cation is a positively charged atomic or molecular particle.

■ cation exchange

a cationic resin has positive ions built into its structure and therefore exchanges negative ions. In cation exchange, the side groups are ionised acidic groups, such as ($-SO_3H$, -COOH, -OH) to which cations H^+ are attached. The exchange reaction is one in which different cations in the solution displace the H^+ from the solid.

Cation Exchange Illustrated

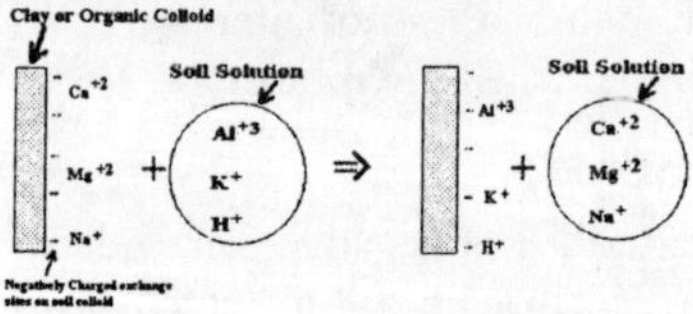

■ cation-exchange chromatography

the form of ion exchange chromatography that uses resins or packings with functional groups that can separate cations. A sulphonic acid would be an example of a strong cation-exchange group, a carboxylic acid would be a weak cation-exchange group.

■ cationic polymerisation

an ionic polymerisation in which the kinetic-chain carriers are cations.

■ cationic surfactant

a surfactant with a positively charged ionic group. The most common cationic surfactants are known as quaternary ammonium compounds such as alkyl dimethyl benzyl ammonium chloride. These are widely used as disinfectants and sanitisers.

■ caustic soda

(sodium hydroxide) A common reagent that is widely used in the manufacture of soap.

■ CCR

conradson carbon residue.

■ cement

a substance that can be used to build together aggregates of sand or stone into a cohesive structure. May be a single compound or a mixture. May be hydraulic set, air set or chemical set.

■ cementite

iron carbide (Fe_3C).

■ centigrade

temperature based on 0 for the temperature at which water freezes and 100 for the temperature at which water boils.

■ centipoise (cp)

0.01 poise or centistokes times specific gravity at the test temperature.

■ centistoke (cst)

see **viscosity**.

■ centrifuge

a machine using centrifugal force produced by high-speed rotation for separating materials of different densities. Applied to Diesel engine fuels and lubricating oils to remove moisture and other extraneous materials.

■ ceramic

1. of or relating to products, such as pottery, porcelain, or tile, that are made from nonmetallic mineral.
2. an object made of such a material. an object made of such a material.

■ ceramic coating

1. the process of applying a layer of ceramic material, such as alumina, onto a metallic product in order to protect against extremely high temperatures.
2. the material thus applied.

■ ceramic tile

clay tile with an impervious, usually glossy, layer on the surface.

■ ceramics

ceramics are an inorganic material of very high melting point. Ceramics are metal silicates, oxides, nitrides, etc.

■ cermet

a composite material consisting of a combination of ceramic and metallic materials.

■ cetane index

an empirical measure of ignition quality. Defined as the percentage by volume of cetane in a mixture of cetane and methyl naphthalene which has the same ignition quality when used in an engine as a fuel under test.

■ cfr diesel fuel testing unit

a standard engine employed in making cetane number tests of Diesel engine fuels.

■ cgs system of units

cgs system of units is a system of units based upon the centimetre, gram, and second. The International System (SI) has supplanted the cgs system.

■ chain

the whole or part of a macromolecule, an oligomer molecule, or a block, comprising a linear or branched sequence of constitutional units between two boundary constitutional units, each of which may be either an end-group, a branch point, or an otherwise-designated characteristic feature of the macromolecule.

chain length

the length of carbon chain in the hydrocarbon of a reversed-phase packing. It is expressed as the number of carbon atoms (e.g., C_8. C_{18}).

chain polymerisation

a chain reaction in which the growth of a polymer chain proceeds exclusively by reaction(s) between monomer(s) and reactive site(s) on the polymer chain with regeneration of the reactive site(s) at the end of each growth step.

1). A chain polymerisation consists of initiation and propagation reactions, and may also include termination and chain transfer reactions.

2). The adjective *chain* in *chain polymerisation* denotes a chain reaction rather than a polymer chain.

3). Propagation in chain polymerisation usually occurs without the formation of small molecules. However, cases exist where a low-molar-mass by-product is formed, as in the polymerisation of oxazolidine-2,5-diones derived from amino acids (commonly termed amino-acid *N*-carboxy anhydrides). When a low-molar-mass by-product is formed, the adjective *condensative* is recommended to give the term condensative chain polymerisation.

5). The term *chain polymerisation* may be qualified further, if necessary, to specify the type of chemical reactions involved in the growth step, e.g., ring-opening chain polymerisation, cationic chain polymerisation.

6). There exist, exceptionally, some polymerisations that proceed *via* chain reactions that, according to the definition, are not chain polymerisations. The classification can be made more precise and the polymerisation described as a chain-reaction polyaddition.

chain reaction

a reaction where the product of one step is a reactant in a later step, which produces a reactant for a later step, and so on.

chain scission

a chemical reaction resulting in the breaking of skeletal bonds .

chalcogens

chalcogens are the Group 16 elements: oxygen (O), sulphur (S), selenium Se), tellurium (Te), and polonium (Po). Compounds of these elements are called chalcogenides.

chalking

formation of a friable powder on the surface of a paint film caused by the disintegration of the binding medium due to disruptive factors during weathering. The chalking of a paint film can be considerably affected by the choice and concentration of the pigment. It can also be affected by the choice of the binding medium.

channeling

1.occurs when voids created in the packing material of a column may cause mobile phase and accompanying solutes to move more rapidly than the average flow velocity, resulting in band broadening.

The voids are created by poor packing or by erosion of the packed bed.
2. formation of a 'groove' in grease (or in oil too viscous to flow readily under existing conditions). Channels are cut by the motion of a lubricated element, such as a gear or the rolling member of an anti-friction bearing. The amount of channeling can be controlled to a large extent by the consistency or viscosity of the lubricant. While some degree of channeling is desirable to prevent excessive churning of the lubricant, particularly in high speed rolling element bearings, a channel so permanent as to preclude further movement of lubricant to the contacting surfaces might cause equipment failure due to lack of lubricant.

■ characteristic curve

a curve in which D is plotted against Log E resulting from photographic plates exposed to a constant light I (power intensity) for a series of time(time-scale exposures).

■ Charles' law

the volume of a fixed mass of gas at constant pressure expand by constant fraction of its volume at 0 °C for each Celsius degree or Kelvin its temperature is raised. For any ideal gas fraction is approximately 1/273.
This is equivalent to the statement that the volume of a fixed mass of gas at constant pressure is proportional to its thermodynamic temperature

$V = kT$

Charles's Law

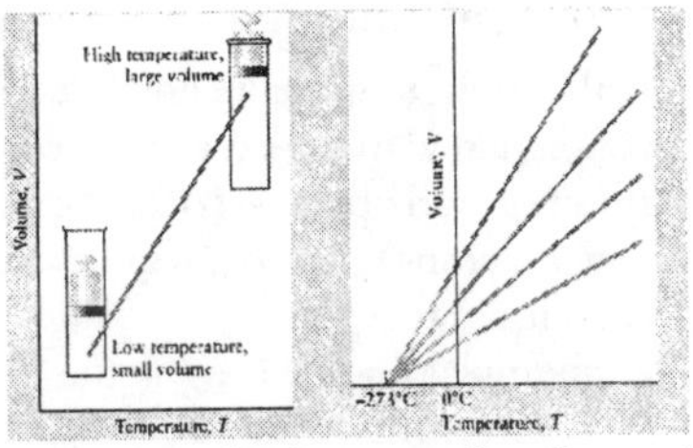

This law also know as Gay-Lussac's law.
An equation similar to that given above applies to pressures for ideal gases.

■ chatter

the sawtooth pattern along a cut line on artwork patterns caused by an uneven blade application while scribing the line cut.

■ chelate

chelate is a compound characterised by the presence of bonds from two or more bonding sites within the same ligand to a central metal atom.

■ chelating agent

an organic sequestering agent used to inactivate hard water and other metallic ions in water. Additives in detergents for inactivating the minerals in water that interfere with cleaning. Ingredients include ethylene diamine tetraacetic acid (EDTA), NTA and sodium citrate.

■ chelatory agent

a substance that binds minerals to itself. In cosmetic manufacture it refers to chemicals that mop up free ions, such as metals, in formulae and inhibit them from causing deterioration of the product.

chemical engineering

1. the branch of engineering that is concerned with the design and construction and operation of the plants and machinery used in industrial chemical processes.
2. chemical engineering is the use of science, mathematics, and economics to convert starting materials or chemicals into more useful forms. We do this through operations called chemical processes, which often consist of many separate steps. Such chemical processing results in thousands of products that are part of virtually every aspect of our lives. Chemical engineers use the principles of chemistry, mathematics, and physics to design and operate the various steps that make up complex chemical processes.

chemical etching

selective stock removal from the base by chem etch, etching, chemmachining or conversion of material by electrochemical reaction.

chemical milling

selective stock removal from the base by chem etch, milling, chemmachining or conversion of material by electrochemical reaction.

chemical potential

for a mixture of substances, the chemical potential of constituent B (i_B) is defined as the partial derivative of the Gibbs energy *G* with respect to the amount (number of moles) of B, with temperature, pressure, and amounts of all other constituents held constant. Also called partial molar Gibbs energy. Components are in equilibrium if their chemical potentials are equal.

chemical set cement

a cement that sets through reaction or precipitation. Often subjected to a high temperature during manufacture or use.

chemisorption

sorption caused by a chemical reaction with the packing. Most such interactions are irreversible; usually occur on packings with reactive functional groups such as silanol or bonded amino phases .

chiral molecule

chiral molecule is a molecule which cannot be superimposed on its mirror image. A common example is an organic molecule containing a carbon atom to which four different atoms or groups are attached. Such molecules exhibit optical activity, i.e., they rotate the plane of a polarised light beam.

Chiral Stationary Phases (CSP)

stationary phases that are designed to separate enantiomeric compounds. They can be bonded to solid suppons or created in situ on the surface of the solid suppon, or they can be surface cavities that allow specific interactions with one enantiomeric form.

■ chirality

the 'twist' of a nanotube, measured by the length of two vectors. Each vector passes perpendicularly through the hexagons that make up a nanotube, in two different directions from a single point. If the path of these hexagons does not go around the nanotube, but can be followed straight along the length of the tube, the chirality is (10,10). The chirality is important for determining whether the nanotube is metallic or semiconducting.

■ chlorofluorocarbons (CFC's)

chlorofluorocarbons and many chlorinated solvents such as 1,1,1 Trichloroethane destroy ozone, leaving holes in the Ozone Layer, especially over the poles. These holes are expanding towards the equator, and as a result the levels of harmful UV radiation in areas such as South America are dangerously high. UV radiation causes skin cancer and blindness in humans (and animals) and destroys plant life. Many waterproofing products are sold in aerosol form. Until recently, aerosols nearly all used CFC's as propellants. International laws and public pressure have now forced most manufacturers to switch to ozone friendly propellants, but many still use 1,1,1 Trichloroethane as a solvent to carry their active ingredients onto the garment. This is an extremely ozone-damaging chemical, yet some of these products still claim to be 'Ozone Friendly'.

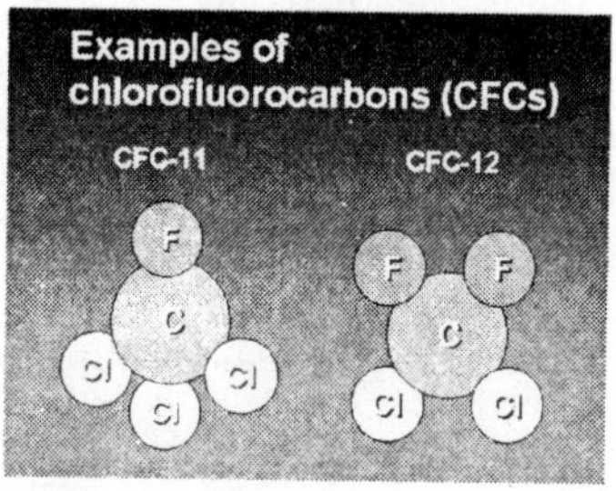

■ chromatogram

a plot of detector signal output versus time or elution volume during the chromatographic process.

■ chromatography

chromatography is a method for separation of the components of a sample in which the components are distributed between two phases, one of which is stationary while the other moves. In gas chromatography, the gas moves over a liquid or solid stationary phase. In liquid chromatography, the liquid mixture moves through another liquid, a solid, or a gel. The mechanism of separation of components may be adsorption, differential solubility, ion-exchange, permeation, or other mechanisms.

■ cis-trans isomerism

compounds with double bonds, or alicyclic rings can exhibit isomerism, due to the attached groups lying above or below the plane of the double bond or ring. The '*cis*' compound is the one with the groups on the same side of the bond, and the '*trans*' has the groups on the opposite sides. The different isomers have different physical and chemical properties. Examples are *fumaric acid/maleic acid* and *1,3-dimethylcyclohexane* - see below:

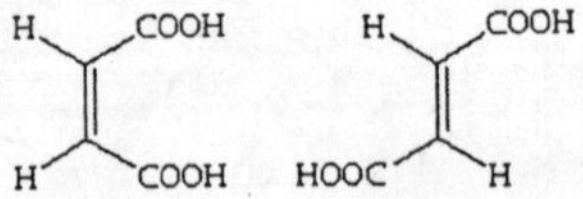

cis- Maleic acid   trans- Fumaric acid

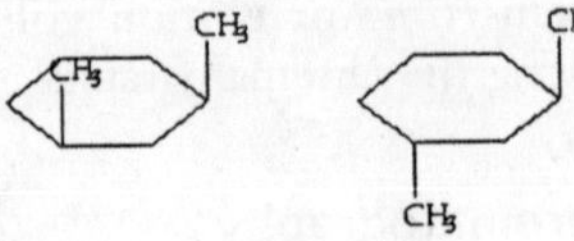

cis-   trans-

■ cladding

the metallic outer sheath of a fuel rod element generally made of a zirconium alloy. It is intended to isolate the fuel from the external environment.

■ Clapeyron equation

Clapeyron equation is a relation between pressure and temperature of two phases of a pure substance that are in equilibrium.

■ clarifier

a machine used for a liquid-sludge separation in which the particles with a higher specific gravity are separated from the lower specific gravity of the liquid. A clarifier bowl has one outlet for the light phase oil; the heavier phase particles are retained on the bowl wall.

■ clay

a rock or mineral fragment of any composition that is smaller than very fine silt grains, having a diameter less than 0.00016 in. (1/256 mm). A clay mineral is one of a complex and loosely defined group of finely crystalline hydrous silicates formed mainly by weathering or alteration of primary silicate minerals. They are characterised by small particle size and their ability to adsorb large amounts of water or ions on the surface of the particles.

■ cleanser

a powdered or liquid cleaning product containing abrasives, surfactants and frequently a bleach.

■ cloning

the act of creating genetically identical cells, tissues, or organisms through the asexual division of genetic material

■ closed loop

controller in automatic mode.

■ cloud point

the temperature at which a surfactant becomes insoluble in water. This becomes important when designing detergents for use in hot water.

■ CNC machining

Computer Numerical Control. Prior to using a computer in controlling machine tools it was simply, nc or numerical or numeric control.

coagulation

an irreversible process in which a number of emulsion droplets coalesce, leading to complete separation of the emulsion.

coat

to cover or apply onto a substrate a photo resist film material coating as by dipping, rolling, spraying, laminating, spinning, or flow.

coating

1. any mineral substance used to cover the surface of a printing paper, such as blanc fixe or china clay.
2. the application of such a substance to a sheet of paper.
3. the application of a varnish or photosensitive solution to a printing plate.

codisposal

a packaging method for disposal of radioactive waste in which more than one type of waste are combined in disposal containers. Codisposal takes advantage of otherwise unused space in disposal containers and is more cost-effective than other methods of limiting the reactivity of individual waste packages.

coemulisifiers

secondary emulsifiers that improve the stability of emulsion systems.

cohesion

1. in geology, the portion of the shear strength of a rock not related to inter-particle friction.
2. in physics, the attraction between molecules of a liquid that allows formation of drops or films of that liquid.

cola-nut

a natural stimulant - provides energy

cold working

the plastic deformation of a metal at a temperature below that at which it recrystallises.

colloid

colloids are systems in which there are two or more phases, with one (the dispersed phase) distributed in the other (the continuous phase). Moreover, at least one of the phases has small dimensions, in the range between 1 nanometer and 1 micrometer (10^{-9} - 10^{-6} m). Dimension, rather than the nature of the material, is characteristic. In this size range, the surface area of the particle is large with respect to its volume so that unusual phenomena occur, e.g., the particles do not settle out of the suspension by gravity and are small enough to pass through filter membranes. Macromolecules (proteins and other high polymers) are at the lower limit of this range; the upper limit is usually taken to be the point at which the particles can be resolved in an optical microscope. Colloidal particles may be gaseous, liquid, or solid, and occur in various types of suspensions:
Sols - dispersions of small solid particles in a liquid.

Emulsions - colloidal systems in which the dispersed and continuous phases are both liquids.
Gels - colloids in which both dispersed and continuous phases have a three-dimensional network throughout the material.
Aerosols - colloidal dispersions of liquid or solid particles in a gas.
Foams - dispersions of gases in liquids or solids.

■ colloid mills

colloid mills are machines used to grind aggregates into very fine particles or to apply very high shearing within a fluid to produce colloid suspensions or emulsions in which the particle sizes are less than 1 micrometer. One type of colloid mill is called a disc mill, in which a mixture of a solid and liquid (or two liquid) is passed between two discs a small distance apart, which rotate very rapidly relative to each other. Applications of colloid mills occur in food processing, in paint manufacture, and in the pharmaceutical industry.

■ colloidal instability

a condition of clouds in which the particles tend to aggregate into masses large enough to precipitate.

■ colloidal particle

any of the individual particles of a colloid.

■ colloidal silica

a colloid of fine particles of silica that are negatively charged by a small addition of alkali, forming a high-concentration solution that is used to make molded ceramics or to treat textiles.

■ colloidal suspension

a mixture of two substances, one of which (the colloid) is uniformly distributed throughout the other (the dispersion medium); either substance may be gaseous, liquid, or solid.

■ colluvium

any loose, heterogeneous sediment deposited by rainwash, sheetwash, or slow continuous downslope creep, usually at the base of a cliff or slope.

■ column back pressure

see **head pressure.**

■ column chromatography

any form of chromatography that uses a column or tube to hold the stationary phase. Open-column chromatography, HPLC, and open-tubular capillary chromatography are all examples.

■ column performance

refers to the efficiency of a column; measured as the number of theoretical plates for a given test compound.

■ column switching

the use of multiple columns connected by switching valves to effect better chromatographic separations or for sample cleanup. Fractions from a primary column can be switched to two or more secondary columns, which in turn can be further diverted to additional columns or to the detector(s) .

■ **comb macromolecule**

a macromolecule comprising a main chain with multiple trifunctional branch points from each of which a linear side-chain emanates.

1. If the subchains between the branch points of the main chain and the terminal subchains of the main chain are identical with respect to constitution and degree of polymerisation, *and* the side chains are identical with respect to constitution and degree of polymerisation , the macromolecule is termed a regular comb macromolecule.

2. If at least some of the branch points are of functionality greater than three, the macromolecule may be termed a brush macromolecule.

■ **comb polymer**

a polymer composed of comb macromolecules

■ **complexation capability**

the ability of a chemical element to unite with other elements to form a complex compound.

■ **component model**

the analysis models that are run separately and then combined into process models for running in the TSPA computer model.

■ **composite**

a material brought about by combining materials differing in composition or form on a macroscale for the purpose of obtaining specific characteristics and properties. The constituents retain their identity such that they can be physically identified and they exhibit an interface between one another.

■ **composition**

a process variable. Represents the amount of one material in a solution, or gas.

■ **compound**

a combination of two or more elements, bonded together in some way. It has different physical and chemical properties from the elements it is made of. Compounds are often difficult to split into their elements and can only be separated by chemical reactions.

■ **compounded oil**

a blend of petroleum oil with small amounts of fatty or synthetic fatty oils is referred to as compounding. Compounded oils are used for certain wet applications to prevent washing-off of the lubrication from the metal surfaces. The fatty materials enable the oil to combine physically with the water instead of being displaced. Cylinder oils for wet steam applications and for some air compressors are compounded. Because the fatty materials impart a strong affinity for metal surfaces, compounded oils are frequently used for applications in which lubricity or extra load-carrying ability are needed. They are not generally recommended for service that requires high oxidation stability.

■ **compressible**

a fluid flow is compressible if its density changes appreciably (typically by a few percent) within the domain of interest. Typically, this will occur when the fluid velocity exceeds Mach 0.3. Hence, low velocity flows (both gas and liquids) behave incompressibly.

■ **concentration**

concentration is the amount of given substance in a stated unit of a mixture, solution, or ore. The concentration of an atom, ion, or molecule in a solution may be symbolised by the use of square brackets, as $[Ca^{2+}]$.

■ **conceptual model**

a set of qualitative assumptions used to describe a system or subsystem for a given purpose.

■ **concrete**

a composite material consisting of aggregate particles bound together in a solid body by a cement.

■ **condensation polymerisation**

the formation of polymers by an intermolecular reaction involving at least two monomer species, usually with the production of a low molecular weight byproduct such as water.

■ **condensation reactions**

this is mainly an organic reaction where a simple molecule such as *water* or *ammonia* is eliminated when two molecules combine to produce another compound. A typical reaction is the esterification of an *acid* by an *alcohol* - in this case *water* is eliminated.

■ **CR**

Condensation reactions play a major part in the polymer industry for the synthesis of artificial polymers.

■ **condensed phase explosion**

an explosion that occurs when the fuel is present in the form of a liquid or solid.

■ **conduction**

the flow of thermal energy through a material. Conduction is affected by the amount of heat energy present, the nature of the heat carrier in the material (its thermal conductivity), and the amount of dissipation.

■ **conduction band**

the lowest-lying electron energy band that is not completely filled with electrons.

■ **conductor**

a substance or medium that conducts heat, light, sound or especially, an electric charge.

■ **confidence**

1. in statistics, a measure of how close the estimated value of a random variable is to its true value.

2. degree of assurance that an argument, such as a safety case, is correct.

■ **confidence interval**

an interval that is believed, with a preassigned degree of confidence, to include the particular value of the random variable that is estimated.

configurational base unit

a constitutional repeating unit in a regular macromolecule , a regular oligomer molecule, a regular block, or a regular chain, the configuration of which is defined at least at one site of stereoisomerism in the main chain.

configurational repeating unit

the smallest set of successive configurational base units that prescribes configurational repetition at one or more sites of stereoisomerism in the main chain of a regular macromolecule, a regular oligomer molecule, a regular block, or a regular chain.

configurational sequence

the whole or part of a chain comprising one or more species of configurational unit(s) in defined as sequence.

configurational unit

a constitutional unit having at least one site of defined stereoisomerism.

confined explosion

an explosion of a fuel-oxidant mixture (usually in the gas phase) inside a closed system (e.g. vessel or building).

conformer

a conformer is a molecule which differs from another molecule in that part of it is twisted along a single bond. In other respects, they are exactly the same molecule.

congruent transformation

a transformation of one phase to another that does not involve any change in composition.

consequences

a description of what would occur in the event of a hazard occurring.

conservative assumption

1. an assumption that results in a calculated release of radionuclides exceeding actual or expected releases.
2. an assumption that uses uncertain inputs and does not attempt to include any potentially beneficial effects.

Constitutional Repeating Unit (CRU)

the smallest constitutional unit, the repetition of which constitutes a regular macromolecule, a regular oligomer molecule, a regular block, or a regular chain.

constitutional sequence

the whole or part of a chain comprising one or more species of constitutional unit(s) in defined as sequence.

constitutional unit

an atom or group of atoms (with pendant atoms or groups, if any) comprising a part of the essential structure of a macromolecule, an oligomer molecule, a block, or a chain.

contact printing

a photographic process, with images being transferred from one surface by light exposure to the photosense side of a second. The printed image orientation is dependent on the relative positioning of the surfaces; the tone is dependent on the photosensitive chemical matter used.

■ **containment**

1. the confinement of radioactive waste within a designated boundary.
2. the use of design features to contain or reduce radioactive releases or radiation doses.

■ **continuous reactor**

a reactor that is characterised by a continuous flow of reactants into and out of the reaction system.

■ **continuum model**

a model that represents fluid flow through numerous individual fractures and matrix blocks by approximating them as continuous flow fields.

■ **controlled surface porosity support**

same as Porous-layer bead

■ **convection**

1. the transfer of heat by the circulation of a fluid (either water or air).
2. the bulk motion of a flowing fluid (gas or liquid) caused by temperature differences that, in turn, causes different areas of the fluid to have different densities (e.g., warmer is less dense).

■ **co-oligomer**

an oligomer derived from more than one species of monomer.

■ **co-oligomerisation**

oligomerisation in which a co-oligomer is formed.

■ **coordination number**

the number of atomic or ionic nearest neighbours.

■ **coordinatograph**

an X and Y coordinate plotting machine consisting of a fixed or rotating table & movable head on fixed ways that is capable of precisely locating a point on a line or surface. A coordinatograph is commonly used with cut-and-strip scribe coat materials to generate artwork patterns.

■ **copolymer**

copolymers are also known as heteropolymers. They are made from two (or more) different monomers, which usually undergo a condensation reaction with the elimination of a simple molecule, such as ammonia or water. A typical example is the condensation of 1,6-diaminohexane (hexamethylenediamine) with hexanedioic acid (adipic acid) to form nylon 6,6. The properties of a polymeric plastic can most easily be modified if it is a copolymer of two or more different monomers, e.g. acrylonitrile-butadiene-styrene copolymer (ABS). Varying the proportions of the component monomers can preselect its properties.

■ **copolymerisation**

polymerisation in which a copolymer is formed.

■ copper strip corrosion

this is an evaluation of a product's tendency to corrode copper or copper alloys.

■ copy camera

the cameras adapted for enlarging, reducing, colour separating, & photographic materials screening. Also referred to as a process camera.

■ cordial

a stimulating medicine or drink.

■ corner frequency

for first order time constants, the 'corner frequency' is the frequency where the amplitude ratio starts to turn and the phase lag equals 45 degrees. Also corner frequency = 1/(time constant) radians/time

■ corrosion

deteriorative loss of a metal as a result of dissolution environmental reactions.

■ corrosion inhibitor

1. a material that protects against the wearing away of surfaces. Sodium silicate is a corrosion inhibitor commonly used in detergents. 2. a lubricant additive for protecting surfaces against chemical attack from contaminants in the lubricant. The most common types of corrosion inhibitors generally react chemically with the metal surfaces to be protected, thus forming an inert film in these areas.

■ corrosion resistant material

a material that develops a protective film on its surface, creating a high resistance to corrosion. Such a material, the nickel-based alloy, Alloy 22, would be used as the outer barrier of the two-layer waste package.

■ corrosion

detrimental change in the size or characteristics of material under conditions of exposure or use. It usually results from chemical action either regularly and slowly, as in rusting (oxidation), or rapidly, as in metal pickling.

■ corrosive

a corrosive material is one which causes damage to skin, eyes or other parts of the body on contact.

■ cosine law

a illumination law which indicates that the flux radiated or received in a given direction varies with the projected area of the receiver or emitter in a plane perpendicular to the flux direction.

■ cosmic rays

cosmic rays are high energy (10^{15} - 10^{17} eV) nuclear particles, electrons, and photons, originating mostly outside the solar system, which continually bombard the Earth's atmosphere.

■ Coulomb (C)

the SI unit of electric charge. It is equal to the charge transferred by a current of one ampere in one second (C = A s). The unit is named after French physicist C. A. Coulomb (1736-1806).

■ **Coulomb's law**

Coulomb's law is the statement that the force *F* between two electrical charges q_1 and q_2 separated by a distance *r* is

$$F = \frac{å_0 q_1 q_2}{r^2}$$

where $å_0$ is the permittivity of a vacuum, equal to

$å_0 = 8.8542 \times 10^{-12}$ F/m.

■ **counterion**

in an ion-exchange process, the ion in solution is used to displace the ion of interest from the ionic site. In ion pairing, it is the ion of opposite charge added to the mobile phase to form a neutral ion pair in solution.

■ **coupled columns**

a form of column switching. Uses a primary column connected to two secondary columns vi: a selector valve. Fractions fron the first column can be selectively transferred to the other two columns for additional separation. Term also used to describe two or more columns connected in series to provide increased plate numbers.

■ **coupling agent**

a material used to increase the solubility of one ingredient in another.

■ **covalent bond**

covalent bond is a chemical bond between two atoms whose stability results from the sharing of two electrons, one from each atom.

■ **coverage**

refers to the amount of bonded phase on a silica support in bonded-phase chromatography. Coverage is usually described in ~mol/m or in terms of %C.

■ **cracked**

refers to a petroleum product produced by a secondary refining process such as thermal cracking or vis-breaking processes which yield very low quality residue.

■ **cracking**

breaks or splits in the paint's surface.

■ **crease**

a line, groove, or ridge that is made as if by folding a pliable substance.

■ **creep**

the time-dependent permanent deformation that occurs under stress; for most materials it is important only at elevated temperatures.

■ **critical mass**

critical mass is the minimum mass of a fissionable material (^{235}U or ^{239}Pu) that will initiate an uncontrolled chain reaction as in an atomic bomb. The critical mass of pure ^{239}Pu is about 4.5 kg, and of ^{235}U about 15 kg.

■ **critical micelle concentration**

the concentration of a surfactant in solution at which the molecules begin to form aggregates called micelles while the concentration of surfactant in solution remains constant.

■ critical point

in general, critical point is the point on the phase diagram of a two-phase system at which the two coexisting phases have identical properties and therefore represent a single phase. At the liquid-gas critical point of a pure substance, the distinction between liquid and gas vanishes, and the vapour pressure curve ends. The coordinates of this point are called the critical temperature and critical pressure. Above the critical temperature, it is not possible to liquefy the substance.

■ critical point drying

a procedure used in preparing a specimen for electron microscopic examination, in which damage to the specimen during its drying step is minimised by avoiding exposure to a liquid-gas boundary.

■ critical pressure

critical pressure is the pressure of a fluid in its critical point; i.e. when it is at its critical temperature and critical volume.

■ critical temperature

critical temperature is the temperature of the liquid-vapour critical point, that is, the temperature above which a gas cannot be liquefied by an increase of pressure.

■ critical volume

critical volume is the volume of a fixed mass of a fluid at critical temperature and pressure.

■ criticality

the condition in which nuclear fuel sustains a chain reaction. It occurs when the effective neutron multiplication factor (the number of fissions in one generation divided by the number of fissions in the preceding generation) of a system equals one.

■ cross-drift

an excavation above and across the block of the potential repository excavated in a general northeast–southwest direction.

■ crosslink

a small region in a macromolecule from which at least four chains emanate, and formed by reactions involving sites or groups on existing macromolecules or by interactions between existing macromolecules. The small region may be an atom, a group of atoms, or a number of branch points connected by bonds, groups of atoms, or oligomeric chains. In the majority of cases, a crosslink is a covalent structure but the term is also used to describe sites of weaker chemical interactions, portions of crystallites, and even physical interactions and entanglements.

■ crosslinked polymer

a polymer in which adjacent linear molecular chains are joined at various positions by covalent bonds.

■ cross-linking

cross-linking is attachment of two chains of polymer molecules by bridges, composed of either an element, a group, or a compound, that join certain carbon atoms of the chains by primary chemical bonds.

■ **cross-links**

to increase the rigidity and density of synthetic polymers groups are introduced into the monomer molecules which enable bonds to form between polymer chains - this is known as cross-linking. The degree of cross-linking can have a marked effect on the physical properties of the final product. An example are the *silicones* - they can vary from oils (low degree of cross-linking) to waxes (high degree of cross-linking).

■ **crust**

crust is the outer layer of the solid earth, above the Mohorovicic discontinuity. Its thickness averages about 35 km on the continents and about 7 km below the ocean floor.

■ **cryoscopic constant**

cryoscopic constant (E_f) is the constant that expresses the amount by which the freezing point T_f of a solvent is lowered by a non-dissociating solute, through the relation

$ÄT_f = E_f\, m$

where m is the molality of the solute.

■ **crystal**

a discrete solid where the atoms,ions or molecules are arranged in an ordered 3-dimensional structure. The regular structures have the ability to 'bend' or diffract beams of X-rays, and this aspect of crystallography is an extremely helpful tool in structure determination.

■ **crystal structure**

for crystalline materials, the manner in which atoms or ions are arrayed in space. It is defined in terms of the unit cell geometry and the atom positions within the cell.

■ **crystal system**

a scheme by which crystal structures are classified according to unit cell geometry.

■ **crystalline**

the state of a solid material characterised by a periodic and repeating three-dimensional arrays of atoms, ions, or molecules.

■ **crystallinity**

for polymers, the state wherein a periodic and repeating atomic arrangement is achieved by molecular chain alignment.

■ **crystallite**

a region within a crystalline polymer in which all the molecular chains are ordered and aligned.

■ **CSTR**

continuous stirred tank reactor. This may be thought of as a tank to which reactants flow in, and products flow out. In an ideal CSTR the contents of the reactor are uniformly distributed.

■ **c-term**

the mass transfer term of the van Deemter equation.

■ **cumulative distribution**

for grouped data, a distribution that shows how many of the values are less than or more than specified values. For random variables, this term is synonymous with distribution function.

■ **cumulative probability**

the probability that a random variable will have a value equal to or less than some specified value.

■ **curie**

a unit of radioactivity equal to 37 billion disintegrations per second, abbreviated Ci.

■ **curie temperature**

for a ferromagnetic material, Curie temperature or Curie point (T_C) is the critical temperature above which the material becomes paramagnetic. For iron the Curie point is 760 °C and for nickel 356 °C. It is named after French physicist Pierre Curie (1859-1906).

■ **current density**

amount of electrical current per cross-sectional area of wire.

■ **cut-and-strip:**

a method for producing artwork by cutting the pattern then stripping away the unwanted areas of two layers of material. The engineering terminology 'cut-and-strip' & 'Cut and Peel' are synonymous. The preferable term is 'Cut and Strip.'

■ **cutter stock**

flux Stock. A petroleum stock which is used to reduce the viscosity of a heavier residual stock by dilution.

■ **cyano phase**

a stationary phase that usually consists of cyanopropylsilyl groups; used in both normal and reversed-phase chromatography.

■ **cyclic compounds**

a cyclic compound is one where the atoms are joined together so as to form a closed ring.

■ **cycloalkanes**

cycloalkanes are cyclic saturated hydrocarbons containing a ring of carbon atoms joined by single bonds. They have the general formula C_nH_{2n}, for example cyclohexane, C_6H_{12}. In general, they behave like the alkanes but are rather less reactive.

■ **cyclopolymerisation**

a polymerisation in which the number of cyclic structures in the constitutional units of the resulting macromolecules is larger than in the monomer molecules .

■ **dead time**

dead time is the amount of time that it takes for your process variable to start changing after your valve changes. If you were taking a shower, the dead time is the amount of time it would take for you (the controller) to feel a change in temperature after you have adjusted the hot or cold water.

Pure dead time processes are usually found in plug flow or solids transportation loops. Examples are paper machine and conveyor belt loops. Dead time is also called delay. A controller cannot make the process variable respond before the process dead time.

To a controller, a process may appear to have more dead time than what it actually has. That is, the controller cannot be tuned tight enough (without going unstable) to make the process variable respond appreciably before an equivalent dead time. More accurately, the characteristic time of the loop is determined by equivalent dead time. Equivalent dead time consists of pure dead time plus process components contributing more than 180 degrees of phase lag.

The phase of dead time increases proportionally with frequency. Any process having more than 180 degrees phase lag has equivalent dead time.

■ dead volume (v_d)

the volume outside of the column packing itself. The interstitial volume (intraparticle volume + interparticle volume) plus extracolumnvolume (contributed by injector, detector, connecting tubing, and endfittings) all combine to create the dead volume. This volume can be determined by injecting an inen compound (i.e.. a compound that does not interact with the column packing). Also abbreviated V_0 or V_m.

■ decomposition

the breaking up of a chemical compound into by-products. The temperature at which decomposition is observed depends on scale and is markedly dependent on the sensitivity of the measuring equipment.

■ decomposition energy

the decomposition energy is the maximum amount of energy which can be released upon decomposition. The product of decomposition energy and total mass is an important parameter for determining the effects of a sudden energy release, e.g. in an explosion. The decomposition energy can occasionally be obtained from literature or calculated theoretically, however it is mostly determined experimentally, often by DSC

■ decomposition temperature

temperature at which spontaneous decomposition occurs. In practice, it is impossible to indicate the exact value of this temperature, because the reaction rate is only zero at absolute zero temperature (0 K) conforming to the equation of Arrhenius. Therefore, in measuring the decomposition temperature both sample quantity and sensitivity of the measuring device are very important.

■ decovalex

an international software code testing program.

■ defence in depth

1. a design strategy based on a system of multiple, independent, and redundant barriers, designed to ensure that failure in any one barrier does not result in failure of the entire system.

2. a system of multiple barriers that mitigate uncertainties in conditions, processes, and events.

■ **definition**

the fidelity of reproduction with the pattern edge relative to the original engineering master.

■ **defoamers**

substance used to reduce or eliminate foam.

■ **degassing**

the process of removing dissolved gas from the mobile phase before or during use. Dissolved gas may come out of solution in the detector cell and cause baseline spikes and noise. Dissolved air can affect electrochemical detectors (by reaction) or fluorescence detectors (by quenching). Degassing is carried out by heating the solvent or by vacuum (in a vacuum flask), or on-line using evacuation of a tube made from a gas-permeable substance such as PTFE, or by helium sparging.

■ **degradation**

degradation is the term generally used to describe the loss of resilience of material used for protective gloves. Degradation may cause the material to soften, swell, become hard and brittle, or - in severe cases - disintegrate.

■ **degreaser**

a specialty product that removes grease and greasy/oily soils from hard surfaces. Basic ingredients include surfactants that penetrate and emulsify along with alcohol or glycol derivatives to boost cleaning.

■ **degree of polymerisation**

degree of polymerisation is the number of monomeric units in a macromolecule or an oligomer molecule.

■ **deionised water**

water from which charged or ionisable organic or inorganic salts are removed.

■ **delay**

this term is often used in place of dead time.

■ **deliquescence**

the absorption of water vapour from the air by a crystalline solid, leading to dissolution of the solid.

■ **demulsibility**

the test time required for a specified oil-water emulsion to break. Highly refined straight mineral oils have inherently good demulsibility. Even after violently shaking an oil/water mixture, the oil separates and rises rapidly to the top of the water. This is true also of other oil formulated for good demulsibility. It is a desirable characteristic of oils such as circulating oils that they must separate from water readily. Demulsibility is thus a measure of a lubricating oil's ability to separate from water. This is a important consideration in the maintenance of many circulating oil systems.

■ **density**

in the most common usage, density (*ñ*) is mass density or mass per unit volume.

$\tilde{n} = m / V$

More generally, the amount of some quantity (mass, charge, energy, etc.) divided by a length, area, or volume.

Relative density is the ratio of the density of a substance to the density of some reference substance. For liquids or solids, it is the ratio of the density (usually at 20 °C) to the density of water at 4 °C. This quantity was formerly called specific gravity.

■ density (d)

the log (1/T) where T is transparency. The value for D depends on the photo emulsion component, the magnitude power and nature of the exposure, the processing conditions, and the densitometer optical arrangement.

■ depolymerisation

the process of converting a polymer into a monomer or a mixture of monomers.

■ deposition

the act of depositing.

■ derivative

the 'D' part of PID controllers. With derivative action, the controller output is proportional to the rate of change of the process variable or error. Some manufacturers use the term rate or pre-act instead of derivative. Derivative, rate, and pre-act are the same thing. Derivative action can compensate for a changing process variable. Derivative is the 'icing on the cake' in PID control, and most people don't use it. It can make the controller output jittery on a noisy loop and most people don't use derivative on noisy loops for this reason.

■ desalter

the desalter mixes the hydrocarbon stream with a small amount of fresh water (e.g. 10% by volume) forming a water-in-oil emulsion. The resulting emulsion is subjected to an electric field wherein the water is coalesced as an under flow from the upper flow of a relatively water-free, continuous hydrocarbon phase. The desalted hydrocarbon stream is produced at relatively low cost and has a very small residual salt content. The performance of this unit can be improved with a demulsifier, such as Alken 860 Demulsifier.

■ desalting

technique in which low-molecular-weight salts and other compounds are removed from nonionic and high-molecular-weight compounds. An example is the use of a reversed-phase packing to retain sample compounds by hydrophobic effects but to allow salts to pass through unretained. Use of an SEC column to exclude large molecules and retain lower-molecular-weight salts is another example.

■ design bases

information that identifies the specific functions to be performed by a structure, system, or component of a facility and the specific values or ranges of values chosen for controlling parameters as reference bounds for design. These values may be constraints derived from generally accepted state-of-the-art practices for achieving functional goals or requirements derived from analysis (based on calculation or experiments) of the effects of a postulated event under which a structure, system, or component must meet its functional goals. The values for controlling parameters for external events include 1. estimates of severe natural events to be used for deriving design bases that will be based on consideration of historical data on the associated parameters, physical data, or analysis of upper limits of the physical processes involved; and 2. estimates of severe external human-induced events, to be used for deriving design bases, that will be based on analysis of human activity in the region taking into account the site characteristics and the risks associated with the event.

■ Design of Experiments, DOE

a statistical method used to determine the interrelationships between two or more process variables.

■ desorption

a physical or chemical process by which a substance that has been adsorbed or absorbed by a liquid or solid material is removed from the material.

■ desquamation

desquamation is the detachment of cells from the surface of an epithelium.

■ detonation

a violent explosion involving high-velocity pressure waves; in a gasoline engine, the spontaneous combustion of part of the compresses charge after spark occurs. Detonation usually produces a characteristic metallic sound, or knock

■ develop

to subject photosensitive film material to a chemical treatment designed in producing a usable image from matter previously modified by radiation.

■ devitrification

the conversion of a glassy substance to a crystalline substance; for example, the alteration of glass in vitrified tuff into zeolites.

■ dewar

a jacketed vacuum flask that can be used in calorimetry. The low rate of heat loss means that it can be used to simulate the behaviour of large reactors. See also adiabatic.

■ diamagnetism

in diamagnetism the magnetisation is in the opposite direction to that of applied field, i.e. susceptibility is negative. It results from changes induced in the orbits of electrons in the atoms of a substance by the applied field, the direction of the change opposing the applied flux.

■ diazo compounds

1. diazo compounds are compounds having the divalent diazo group, $=N^{+}=N^{-}$, attached to a carbon atom. The term includes

azo compounds, diazonium compounds, and also such compounds as diazomethane, $CH_2=N_2$.
2. this is a series of organic compounds possessing the general formula **Ar-N=N-Y**, where Ar is an aromatic group and Y is any group linked to the nitrogen except via a carbon atom. They are different from the azo compounds which have the general formula **Ar-N=N-Ar**. Diazo compounds are formed from amines by a two stage diazotisation reaction. Stage one is the conversion of an aromatic amine into a diazonium chloride salt using cold *sodium nitrite* and *hydrochloric acid*.
The second stage is the reaction of the diazonium chloride salt with an appropriate compound - such as another aromatic amine, eg. the reaction with *p-toluidine*.

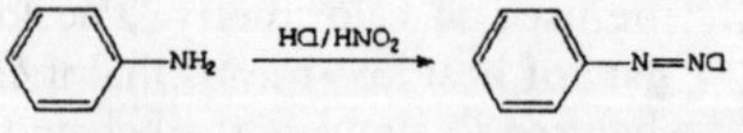

3. a nonsilver, room lighting, UV sensitive coating usually on a stable transparent film substrate. diazo coatings yield mirror image with duplicate tone through contact printing and are developed in ammonia vapour. The image has high actinic density & visual transparency.

■ dielectric

any material that is electrically insulating.

■ dielectric constant

dielectric constant or permittivity (ǻ) is an index of the ability of a substance to attenuate the transmission of an electrostatic force from one charged body to another. The lower the value, the greater the attenuation. The standard measurement apparatus utilises a vacuum whose dielectric constant is 1. In reference to this, various materials interposed between the charged terminal have the following value at 20 °C:
air 1.00058
glass 3
benzene 2.3
acetic acid 6.2
ammonia 15.5
ethanol 25
glycerol 56
water 81
The exceptionally high value for water accounts for its unique behaviour as a solvent and in electrolytic solutions. Dielectric constant values decrease as the temperature rises.

■ dielectric strength

the magnitude of an electric field necessary to cause significant current passage through a dielectric material.

■ dienes

dienes are unsaturated organic compounds that contain two fixed double bonds between carbon atoms. Dienes in which the two double-bond units are linked by one single bond are termed conjugated.

■ diesel index

product of the API gravity and the aniline point (in degrees Fahrenheit) of a Diesel fuel, divided by 100; an indication of the ignition quality of the fuel.

■ **differential scanning calorimetry**

a technique in which the change of the difference in the heat flow rate to the sample and to a reference sample is measured while they are subjected to a temperature regime. Note that in many process safety laboratories DSC and DTA are used interchangeably.

■ **differential thermal analysis**

a technique in which the change of the difference in the temperature between the sample and the reference sample is measured while they are subjected to a temperature regime.

■ **diffuse**

to spread out freely; scatter

■ **diffusion**

diffusion is the spontaneous mixing of one substance with another when in contact or separated by a permeable membrane. Diffusion is a result of the random motions of their component atoms, molecules, ions, or other particles. Diffusion occurs most readily in gases, less so in liquids, and least in solids. The rate of diffusion is proportional to the concentration of the substance and increases with temperature. The theoretical principles are stated in Fick's laws.

■ **diffusion coefficient**

a material's weight, in grams, as it diffuses in 1 second through 1 square centimeter of a medium of a determined concentration gradient (i.e., a medium having a known variability in the concentration of the substance in solution as it travels over distance).

■ **diffusive transport**

movement of molecules or particles due to their concentration gradient. Occurs when dissolved or suspended radionuclides move from regions of higher or lower concentration.

■ **dilatant fluids**

viscosity increases with increasing velocity gradient. They are uncommon, but suspensions of starch and sand behave in this way. Dilatant fluids are also called as shear thickening fluids.

■ **dilution**

the reduction of a dissolved substance's concentration in a solution, caused by an increase in the solvent's proportion of the solution. The solvent frequently is water.

■ **dimethicone**

dimethicone copolyol, a polymer based on silicone, which is used as a conditioner in hair care products and as a skin protectant. It has a tendency to build up on the hair shaft, eventually making the hair appear dull and is not, therefore, currently favoured. It is not considered toxic.

■ **DIMM**

dual in-line memory module.

■ **diodes**

any electronic device that restricts current flow chiefly to one direction. Or a two-terminal semiconductor device used chiefly as a rectifier.

■ diol phase

a stationary phase useful in both normal and reversed-phase chromatography. Consists of a diol structure (two -OH groups on adjacent carbon atoms in an aliphatic chain). Less polar than silica stationary phases used in normal-phase chromatography; has been used for the reversed-phase separation of proteins and polypeptides.

■ dipentene

another name for limonene.

■ dipole

dipole is a pair of separated opposite electric charges. Electric dipole is an assemblage of atoms or subatomic particles having equal electric charges of opposite sign separated by a finite distance. In the case of HCl, the electrons are attracted towards the more electronegative chlorine atom.

■ dipole moment

electric dipole moment (*i*) is the product of the positive charge and the distance between the charges. Dipole moments are often stated in debyes; The SI unit is the coulomb metre. In a diatomic molecule, such as HCl, the dipole moment is a measure of the polar nature of the bond; i.e. the extent to which the average electron charges is displaced towards one atom (in the case of HCl, the electrons are attracted towards the more electronegative chlorine atom). In a polyatomic molecule, the dipole moment is the vector sum of the dipole moments of the individual bonds. In a symmetrical molecule, such as tetrafluoromethane (CF_4) there is no overall dipole moment, although the individual C-F bonds are polar.

■ discrete i/o

senses or sends either 'on or off' signals to the field. For example a discrete input would sense the position of a switch. A discrete output would turn on a pump or light.

■ discrete logic

refers to digital or 'on or off' logic. For example, if the car door is open and the key is in the ignition, then the bell rings.

■ dislocation

a linear crystalline defect around which there is an atomic misalignment.

■ dispersing agent

a material that reduces the cohesive attraction between like particles.

■ dispersion

a process of scattering or separating; specific uses include:
Physical Chemistry. a two-phase system consisting of finely divided particles (the disperse phase) distributed throughout a bulk substance (the continuous phase); for example, fog is a dispersion of liq-

uid particles in a gas; paint is a dispersion of solid particles in a liquid.
Physics. the separation of a complex wave containing several frequencies into its individual component waves by virtue of different wave speeds.
Astrophysics. the slowing down of a radio signal's propagation velocity as it passes through an ionised gas or plasma in the interstellar medium; short frequencies are the least affected and long ones the most affected.
Electromagnetism. the scattering of microwave radiation by an obstruction.
Mineralogy. the characteristic and specific manner in which a monopaque mineral will refract white light.

■ dispersion (hydrodynamic dispersion)

1. the tendency of a solute (substance dissolved in groundwater) to spread out from the path it is expected to follow if only the bulk motion of the flowing fluid moved it.
2. the macroscopic outcome of the actual movement of individual solute particles through a porous medium. Dispersion causes dilution of solutes, including radionuclides, in groundwater and is usually an important mechanism for spreading contaminants in low flow velocity situations.

■ dispersivity

the degree to which small particles of a solid are distributed throughout either a liquid or another solid.

■ displacement chromatography

chromatographic process in which the sample is placed onto the head of the column and is then displaced by a compound that is more strongly sorbed than the compounds of the original mixture. Sample molecules are displaced by each other and by the more strongly sorbed compound. The result is that the eluting sample solute zones may be sharpened; displacement techniques have been used mainly in preparative HPLC appl ications .

■ disposal container

the vessel consisting of the barrier materials and internal components in which the canistered or uncanistered waste form is placed. The disposal container includes the container barriers or shells, spacing structures or baskets, shielding integral to the container, packing contained within the container, and other absorbent materials designed to be placed internal to the container or immediately surrounding the disposal container (i.e., attached to the outer surface of the container). The filled, sealed, and tested disposal container is referred to as the waste package, which is emplaced underground.

■ disruptive processes and events

unlikely processes or events that could affect the performance of the repository, including, for example, human intrusion, volcanic activity, and seismic activity.

■ **disruptive scenario**

a well-defined sequence of events and processes that could adversely affect repository performance initiated by a disruptive process or event.

■ **dissociation**

dissociation is the process by which a chemical combination breaks up into simpler constituents as a result of either added energy (dissociated by heat), or the effect of a solvent on a dissolved polar compound (electrolytic dissociation). It may occur in the gaseous, solid, or liquid state, or in solution.

An example of dissociation is the reversible reaction of hydrogen iodide at high temperatures

$2HI(g)$ à$H_2(g) + I_2(g)$

The term dissociation is also applied to ionisation reactions of acids and bases in water. For example

$HCN + H_2O$ à$H_2O^+ + CN^-$

which is often regarded as a straightforward dissociation into ions

HCN à$H^+ + CN^-$

■ **dissociation constant**

dissociation constant is a constant whose numerical value depends on the equilibrium between the undissociated and dissociated forms of a molecule. A higher value indicates greater dissociation.

The term dissociation is also applied to ionisation reactions of acids and bases in water. For example

$HCN + H_2O$à $H_2O^+ + CN^-$

which is often regarded as a straightforward dissociation into ions

HCN à $H^+ + CN^-$

The equilibrium constant of such a dissociation is called the acid dissociation constant or acidity constant, given by

The concentration of water $[H_2O]$ can be taken as constant. Similarly, for a base, the equilibrium

NH_3 à $NH_4^+ + OH^-$

is also a dissociation; with the base dissociation constant or basicity constant, given by

K_a (K_b) is a measure of the strength of the acid (base).

■ **dissolution**

the dissolving of a solid or gas in a liquid.

■ **distillation**

1. distillation is the process of boiling a liquid and condensing and collecting the vapour. The liquid collected is the distillate. The usual purpose of distillation is purification or separation of the components of a mixture. This is possible because the composition of the vapour is usually different from that of liquid mixture from which it is obtained. Gasoline, kerosene, fuel oil, and lubricating oil are produced from petroleum by distillation.

2. the separation and purification of a mixture of components by vaporisation followed by condensation, based on the different vola-

tilities of each component. A typical experimental set up for distillation is shown below:

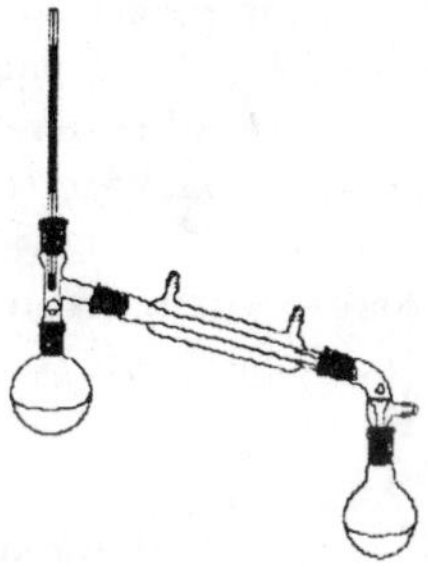

■ **distilled water**

water which has had salts removed by distillation. It is very pure, but does contain some dissolved gases.

■ **distribution**

the overall scatter of values for a set of observed data. A term used synonymously with frequency distribution. Distributions have probability structures that are the probability that a given value occurs in the set.

■ **distribution function**

a function whose values are the probabilities that a random variable assumes a value less than or equal to a specified value. Synonymous with cumulative distribution.

■ **doe spent nuclear fuel**

radioactive waste created by defense activities. The major contributor to this waste form is the N Reactor fuel currently stored at the Hanford Site. This waste form also includes naval spent nuclear fuel.

■ **dominant dead time process**

if the dead time is larger than the lag time the process is a dominant dead time process.

■ **dominant lag process**

most processes consist of both dead time and lag. If the lag time is larger than the dead time, the process is a dominant lag process. Most process plant loops are dominant lag types. This includes most temperature, level, flow and pressure loops.

■ **dopant**

a small quantity of a substance, such as phosphorus, added to another substance, such as a semiconductor, to alter the latter's poperties.

■ **doping**

the intentional alloying of semiconducting materials with controlled concentrations of donor or acceptor impurities.

■ **dose**

the amount of radioactive energy taken into (absorbed by) living tissues.

■ **dose equivalent**

the product of the absorbed dose in tissue, quality factor, and all other necessary modifying factors at the location of interest.

■ **double-strand chain**

a chain that comprises constitutional units connected in such a way that adjacent constitutional units are joined to each other through three or four atoms, two on one side and either one or two on the other side of each constitutional unit.

■ **double-strand copolymer**

a copolymer, the macromolecules of which are doublestrand macromolecules.

■ **double-strand macromolecule**

a macromolecule that comprises constitutional units connected in such a way that adjacent constitutional units are joined to each other through three or four atoms, two on one side and either one or two on the other side of each constitutional unit.

■ **double-strand polymer**

a polymer, the macromolecules of which are doublestrand macromolecules.

■ **downgradient**

the direction that water will tend to flow as the result of a difference in pressure, as indicated by the elevation change in the potentiometric surface.

■ **drawing**

a deformation technique used to fabricate metal wire and tubing. Deformation is accomplished by pulling the material through a die by means of a tensile force applied on the exit side.

■ **drift**

from mining terminology, a horizontal underground passage. Includes excavations for emplacement (emplacement drifts) and access (main drifts).

■ **drip shield**

a corrosion-resistant engineered barrier that is placed above the waste package to prevent seepage water from directly contacting the waste package for thousands of years. The drip shield also offers protection to the waste package from rockfall.

■ **drying**

1. the removal of most of the water from a substance (usually 92-95%), often through heat exposure.
2. the removal of solvents other than water from a substance.

■ **drying agent**

a substance having an affinity for water, used to dry fluid substances; a desiccant.

■ **dual permeability conceptual model**

a conceptual model of groundwater flow in which fractures and rock matrix are represented as separate, interacting continua, with no assumption of pressure equilibrium between fractures and rock matrix. This concept allows modelling groundwater flow as occurring mostly in the fractures, with less flow in the rock matrix, depending on the degree of connection between the rock matrix and fractures and the capillary pressure gradient. The dual permeability model is one of the conceptual models for groundwater and heat flow for fractured, porous media.

ductility

a measure of a material's ability to undergo appreciable plastic deformation before fracture.

durable water repellency (dwr)

all waterproof breathable fabrics used for outdoor clothing are treated with a surface water repellent agent in the textile mill before being made up into garments. This is normally a fluorocarbon (Teflon, Scotchguard etc.). This treatment is necessary to keep water from soaking into the outer fabric and making it look and feel wet, 'wetting out'.

dwell or contact time

describes the time.

dyes

dyes are chemical compounds that can attach themselves to fabrics or surfaces to give them colour. Most dyes are complex organic molecules and need to be resistant to many things such as the weather and the action of detergents. *Indigo*, otained from plants was being used by the Egyptians 5000 years ago, and natural dyes obtained from plants and animal sources are stil used today. In 1856, Perkin developed the first synthetic dye based upon *aniline*.

dye-transfer inhibitors

polymers based on polyvinylpyrrolidone that can surround detached dye particles and restrict their 'freedom of movement.' Dye particles that become detached during the washing process can no longer be reattached to the fabric and discolour it. Dye-transfer inhibitors are an important component in colour-care detergents.

ebullioscopic constant

ebullioscopic constant (E_b) is the constant that expresses the amount by which the boiling point T_b of a solvent is raised by a nondissociating solute, through the relation

$$ÄT_b = E_b\ b$$

where b is the molality of the solute.

ecotoxicology

ecotoxicology is the study of the movement of toxic materials through the environment, and the effect that those materials have upon natural systems, in particular the effects on populations and ecosystems. The 'movement' referred to relates not just to the physical movement of toxic materials, under the influence of groundwater flow for example, but also the movement through plant and animal populations via the food web.

eddy diffusion term

the A-term in the van Deemter equation. It is the contribution to plate height that is due to molecules traveling along different paths through the column, depends on the particle size and geometry of the packing. Also called the multipath term. A = 2 (lambda) d_p.

effective dose equivalent

the sum of the products of the dose equivalent received by specified tissues and the appropriate weighting factors applicable to each tissue.

effective porosity

the fraction of a given medium's porosity available for fluid flow and/or solute storage.

effective theoretical plates (n_{eff})

the true number of plates in a column; corrects theoretical plates for dead volume. Neff = $16(t_R - t_m)^2$, where t_m is the void time.

efficiency *(n or h).*

a measure determined by the number of theoretical plates calculated from the equation shown for *H* .

efflorescence

an encrustation of soluble salts, commonly white, deposited on the surface of coatings, stone, brick, plaster, or mortar; usually caused by salts or free alkalies leached from mortar or adjacent concrete as moisture moves through it.

effluent

same as eluate.

eh

a measure of the state of oxidation of a system. Also known as redox potential or oxidation–reduction potential.

elastic modulus

see **modulus of elasticity.**

elasticity

the ability of paint to expand and contract with the substrate without suffering damage or changes in its appearance. Expansion and contraction are usually caused by temperature fluctuations. Some substrates such as yellow pine expand at different rates depending on the type of their grain. Elasticity is a key to durability. Acrylic binders are noted for their elasticity.

elastin

a protein of elastic tissue, ligaments and arterial wall. It is structurally related to collagen and derived from animals.

elastomer

a polymeric material that may experience large and reversible elastic deformations.

elastomers

elastomers are polymers, which are elastic, that is they move with the fibre like a rubber band expands and contracts. All Nikwax polymers are also elastic. These elastomers provide increased performance.

eleais guineensis

palm kernel oil - emollient and lubricant.

electrical resistivity

electrical resistivity (*ñ*) is electric field strength divided by current density when there is no electromotive force in the conductor. Resistivity is an intrinsic property of a material. For a conductor of uniform cross section with area *A* and length L, and whose resistance is R, the resistivity is given by

$$R = \frac{\tilde{n}d}{A}$$

The SI unit is ohm.

electrochemical series

electrochemical series is a series of chemical elements arranged in order of their standard electrode potentials. The hydrogen electrode

$H^+(aq) + e^-$

is taken as having zero electrode potential. An electrode potential is, by definition, a reduction potential. Elements that have a greater tendency than hydrogen to lose electrons to their solution are taken as electropositive; those that gain electrons from their solution are below hydrogen in the series and are called electronegative.

electrode

a solid electric conductor through which an electric current enters or leaves an electrolytic cell or other medium. It collects or emits electric charge, and is often used for detecting electrical impulses in the brain or muscles or transmitting signals in a bionic device.

electrode potential

electrode potential is defined as the potential of a cell consisting of the electrode in question acting as a cathode and the standard hydrogen electrode acting as an anode. Reduction always takes place at the cathode, and oxidation at the anode. According to the IUPAC convention, the term electrode potential is reserved exclusively to describe half-reactions written as reductions. The sign of the half-cell in question determines the sign of an electrode potential when it is coupled to a standard hydrogen electrode. Electrode potential is defined by measuring the potential relative to a standard hydrogen half cell

$H_2(g)$ à $2H^+(aq) + 2e^-$

The convention is to designate the cell so that the oxidised form is written first. For example

$Pt(s) | H_2(g) | H^+(aq) || Zn^{2+}(aq) | Zn(s)$

The e.m.f. of this cell is

$e.m.f. = E_{right} - E_{left}$

By convention, at $p(H_2) = 101325$ Pa and $a(H^+) = 1.00$, the potential of the standard hydrogen electrode is 0.000 V at all temperatures. As a consequence of this definition, any potential developed in a galvanic cell consisting of a standard hydrogen electrode and some other electrode is attributed entirely to the other electrode

$e.m.f. = E(Zn^{2+}/Zn)$

electrolysis

electrolysis is the decomposition of a substance as a result of passing an electric current between two electrodes immersed in the sample.

electrolyte

a substance (e.g., an acid, base, or salt) that, when dissolved in a suitable solvent (e.g., water) or when fused, conducts electric current by the movement of ions instead of electrons.

electrolytic process

a process that causes the decomposition of a chemical compound by the use of electricity.

■ **electromagnetic spectrum**

the entire range of radiation extending in frequency from approximately 10^23 hertz to 0 hertz or, in corresponding wavelengths, from 10^-13 centimeter to infinity and including, in order of decreasing frequency, cosmic ray photons, gamma rays, x-rays, ultraviolet radiation, visible light, infrared radiation, microwaves, and radio waves. It has many uses due to the speed at which it propagates (the speed of light) and because it can be used to transmit great deals of energy or very little (such as radio waves which are virtually harmless at low amplitude).

■ **electromagnetic waves**

all matter absorbs and emmits radiation covering a broad band of frequencies and wavelengths. This electromagnetic radiation has the velocity of the speed of light (2.998×10^8 ms^{-1}) and arises from the vibrating electric charges in atoms and bonds. The range of wavelengths, also known as the electromagnetic spectrum is shown below:

■ **e-wave**

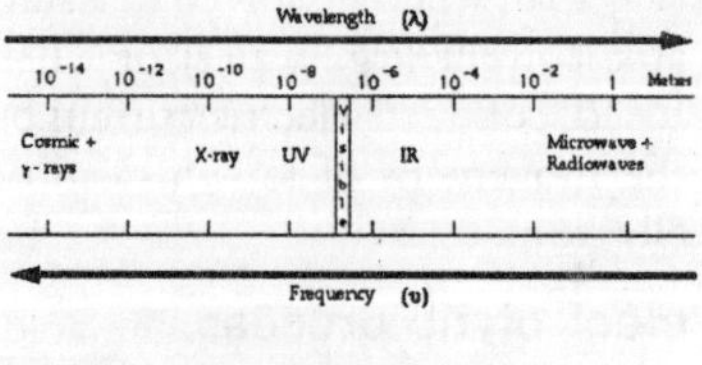

■ **electromotive force**

electromotive force (e.m.f. or E_{MF}) is the difference in electric potential that exists between two dissimilar electrodes immersed in the same electrolyte or otherwise connected by ionic conductors.

■ **electromyography (emg)**

a special recording technique that detects electric activity in the muscle. Migraine patiets are sometimes offered a type of biofeedback called EMG training, in which they learn to control muscle tension in the face, neck and shoulders.

■ **electron**

electron was discovered by J.J. Thompson (1856-1940) in 1896. The electron is an elementary particle with a negative electric charge of $(1.602\ 189\ 2 \pm 0.000\ 004\ 6) \times 10^{-19}$ C and a mass of 1/1837 that of a proton, equivalent to $(9.109\ 534 \pm 0.000\ 047) \times 10^{-31}$ kg. Electrons are arranged in one to seven shells around the nucleus; the maximum number of electrons in each shell is strictly limited by the laws of physics ($2n^2$). The outer shells are not always filled: sodium has two electrons in the first shell ($2\times1^2 = 2$), eight in the second ($2\times2^2 = 8$), and only one in the third ($2\times3^2 = 18$). A single electron in the outer shell may be attracted into an incomplete shell of another element, leaving the original atom with a net positive charge. Valence electrons are those that can be captured by or shared with another atom. Electrons can be removed from the atoms by heat, light, electric energy, or bombardment with high-energy particles. Decaying radioactive nuclei spontaneously emits free electrons, called β(beta) particles.

electron affinity

electron affinity (E_A) is the energy change occurring when an atom or molecule gains an electron to form a negative ion. For an atom or molecule X, it is the energy released for the electron-attachment reaction

$X(g) + e^- X^- (g)$

Often this is measured in electronvolts. Alternatively, the molar enthalpy change, ÄH, can be used.

electron pair

two electrons within one orbital with opposite spins responsible for a chemical bond.

electron photodetachment

the process of detaching an electron from a negative ion using ultraviolet light.

electron spin

electron spin (s) is the quantum number, equal to 1/2, that specifies the intrinsic angular momentum of the electron.

electron volt

electron volt (eV) is a non-SI unit of energy used in atomic and nuclear physics, equal to approximately 1.602177×10^{-19} J. The electron volt is defined as the kinetic energy acquired by an electron upon acceleration through a potential difference of 1 V.

electronegativity

electronegativity is a parameter originally introduced by L. Pauling which describes, on a relative basis, the power of an atom to attract electrons. For example, in hydrogen chloride, the chlorine atom is more electronegative than the hydrogen and the molecule is polar, with negative charge on the chlorine atom. There are various ways of assigning values for the electronegativity of an element. Pauling electronegativities are based on bond dissociation energies using a scale in which fluorine, the most electronegative element, has value 4 and francium, the lowest electronegative element, has value 0.7.

electro-osmotic flow

the overall solvent movement caused by positively-charged ions migrating towards the negative electrode and carrying solvent molecules in the same direction.

electrophoresis

electrophoresis is a technique for the analysis and separation of colloids, based on the movement of charged colloidal particles in an electric field. The migration is toward electrodes of charge opposite to that of the particles. The rate of migration of the particles depends on the field, the charge on the particles, and on other factors, such as the size and shape of the particles.

Electrophoresis is important in the study of proteins. The acidity of the solution can be used to control the direction in which a protein moves upon electrophoresis.

■ **electrostatic attraction**

attractive force between two oppositely charged ions.

■ **elements**

a pure substance that cannot be broken down into smaller substances. Elements are considered the building blocks of all matter. There are just over 100 known elements classified in the periodic table. Elements, compounds and mixtures are the three main types of chemical substances. All substances are made of elements, and most are a combination of two or more elements.

■ **eluate**

combination of mobile phase and solute exiting column; also called effluent.

■ **eluotropic series**

a series of solvents with an increasing degree of polarity, generally used to explain solvent strength in liquid-solid or adsorption chromatography. A nonpolar solvent such as pentane would be at one end of the scale; dichloromethane would oe an intermediate solvent; a strongly polar solvent, such as water, would be at the other end of the scale. Thus, when developing a method or running a gradient, an eluotropic series is useful for selecting solvents.

■ **elution**

the process of passing mobile phase through the column to solutes.

■ **elution chromatography**

the most commonly used chromatographic method. The sample is applied to the head of the column, and individual molecules are separated and eluted at the end of the column. Compare. **displacement chromatography** and **frontal analysis**

■ **elution volume** *(v)*

the volume of mobile phase required to elute a solute from the column at maximum concentration (apex). $V_R = F \cdot t_R$.

■ **elutriation**

a technique used to fractionate packing particles by size. Most ohen used for the separation of ion-exchange resins, such as amino acid resins. that must have a particularly narrow size range. The technique involves the upward flow of water into a large tube. The unsized beads are added to the moving water, and the particles seek their own level, depending on their density and particle size. They are then removed at cenain levels in the tube.

■ **emplacement**

the placement and positioning of waste packages in the repository emplacement drifts.

■ **emplacement horizon**

the area within the repository block where emplacement drifts would be excavated.

■ **emulsification**

the action of breaking up fats, oils and other soils into small particles which are then suspended in a solution.

■ emulsifier

a substance that holds oil in water or water in oil. Examples include borax, beeswax and cetearyl alcohol. Emulsifiers are necessary in the manufacture of creams and lotions, so that they doln't separate.

■ emulsion

emulsion is colloidal system in which the dispersed and continuous phases are both liquids (e.g. oil in water or water in oil). Such systems require an emulsifying agent to stabilise the dispersed particles.

■ emulsion hardening

the process inherent in the developing of layers which renders the desired image abrasion-resistant from handling.

■ enamel

topcoat which is characterised by its ability to form a smooth surface; originally associated with a high gloss, but may also include lower degrees of gloss, i.e., flat enamels. A class of substance having similar composition to glass with the addition of stannic oxide, SnO_2, or other infusible substances to render the enamel opaque.

■ enantiomers

enantiomers are a chiral molecule and its non-superposable mirror image. The two forms rotate the plane of polarised light by equal amounts in opposite directions. Also called optical isomers.

■ endcapping

a column is said to be endcapped when a small silylating agent (e.g., trimethylchlorosilane) is used to bond residual silanol groups on a packing surface. Most often used with reversed-phase packings. May cut down on undesirable adsorption of basic or ionic compounds.

■ endfitting

the fitting at the end of the column that connects it to the injector or detector. Usually made of stainless steel.

■ end-group

a constitutional unit that is an extremity of a macromolecule or oligomer molecule. An end-group is attached to only one constitutional unit of a macromolecule or oligomer molecule.

■ endothermic

a reaction is called endothermic if energy (heat) is absorbed during the reaction

■ energy

energy (*E, U*) is the characteristic of a system that enables it to do work. Like work itself, it is measured in joules (J). The internal energy of a body is the sum of the potential energy and the kinetic energy of its component atoms and molecules. Potential energy is the energy stored in a body or system as a consequence of its position, shape, or state (this includes gravitational energy, electrical energy, nuclear energy, and chemical energy). Kinetic energy is energy of motion and is usually defined as the work that will be done by body possessing the energy when it is brought to rest. For a body of mass *m* having a speed *v*, the kinetic energy is $mv^2/2$. Kinetic energy is most clearly exhibited in gases, in which molecules have much greater freedom of motion than in liquids and solids.

engineered barrier

any component of the engineered barrier system, such as the drip shield, waste package, or invert.

engineered barrier system

the waste packages and the underground facility, including engineered components and systems other than the waste package (e.g., drip shields).

engineering ceramics

technical ceramics for structural applications.

engineering drawing

engineering technical drawing representing the part.

Engler viscosity

a viscosity obtained by dividing the out-flow time in seconds for 200 ml. of the material being tested, by the time in seconds for 200 ml of water at 68F (20C) to flow out of an Engler viscosimeter.

enthalpy

enthalpy (*H*) is a thermodynamic property of a system defined by
$H = U + pV$
where *U* is the internal energy of the system, *p its* pressure, and *V its* volume. J.W. Gibbs put forward the concept of an ensemble in 1902. In a chemical reaction carried out in the atmosphere the pressure remains constant and the enthalpy of reaction (Ä*H*), is to equal
Ä*H* = Ä*U* + *p*Ä*V*
For an exothermic reaction Ä*H* is taken to be negative.

entropy

entropy (*S*) is a measure of the unavailability of a system's energy to do work; in a closed system, an increase in entropy is accompanied by a decrease in energy availability. When a system undergoes a reversible change the entropy (*S*) changes by an amount equal to the energy (*Q*) transferred to the system by heat divided by the thermodynamic temperature (*T*) at which this occurs. All real processes are to a certain extent irreversible changes and in any closed system an irreversible change is always accompanied by an increase in entropy.

enzyme

enzyme is a protein that acts as a catalyst in biochemical reactions. Each enzyme is specific to a particular reaction or group of similar reacuons. Many require the association of certain nonprotein cofactors in order to function. The molecule undergoing reaction (the substrate) binds to a specific active site on the enzyme molecule to form a short-lived intermediate: this greatly increases (by a factor of up to 10^{20}) the rate at which the reaction proceeds to form the product.

■ ep agent

an additive to improve the extreme pressure properties of a lubricant.

■ epoxy resin

epoxy resins are thermosetting resins produced by copolymerising epoxide compounds with phenols (e.g. epichlorohydrin and bisphenol A). They contain ether linkages (-O-) and form a tight, cross-linked polymer network. Toughness, good adhesion, corrosive-chemical resistance, and good dielectric properties characterise epoxy resins. Most epoxy resins are two-part type which harden when blended.

■ eq3/6

a computer software code used to estimate equilibrium mineral phases based on thermodynamic equilibrium, thermodynamic disequilibrium, and reaction kinetics.

■ equation of state

equation of state is an equation relating the pressure, volume, and temperature of a substance or system. Equation of state for ideal gas

$pV = nRT$

where p is pressure, V molar volume, T temperature, and R the molar gas constant (8.314 $JK^{-1}mol^{-1}$).

■ equilibrium constant

the equilibrium constant (K) was originally introduced in 1863 by Norse chemists C.M. Guldberg and P. Waage using the law of mass action. For a reversible chemical reaction represented by the equation

$a\text{A} + b\text{B} \rightarrow c\text{C} + d\text{D}$

chemical equilibrium occurs when the rate of the forward reaction equals the rate of the back reaction, so that the concentrations of products and reactants reach steady-state values. The equilibrium constant is the ratio of chemical activities of the species A, B, C, and D at equilibrium. To a certain approximation, the activities can be replaced by concentrations.

For gas reactions, partial pressures are used rather than concentrations.

The units of K_p and K_c depend on the numbers of molecules appearing in the stoichiometric equation (a, b, c, and d). The value equilibrium constant depends on the temperature. If the forward reaction is exothermic, the equilibrium constant decreases as the temperature rises. The equilibrium constant shows the position of equilibrium. A low value of K indicates that [C] and [D] are small compared to [A] and [B]; i.e. that the back reaction predominates. The equilibrium constant is related to $\Delta_r G°$, the standard Gibbs free energy change in the reaction, by

$RT \ln K = -\Delta_r G°$

■ equivalent continuum model

a conceptual model of groundwater and heat flow, in which the fractures and matrix can be treated as a single composite material, and the hydraulic properties are a combined effect of both fracture and matrix properties.

equivalent dead time

to a controller, a process may appear to have more dead time than what it actually has. That is, the controller cannot be tuned tight enough (without going unstable) to make the process variable respond appreciably before an equivalent dead time. More accurately, the characteristic time of the loop is determined by equivalent dead time consisting of pure dead time plus process components contributing more than 180 degrees of phase lag.

The phase of dead time increases proportionally with frequency. Any process having more than 180 degrees phase lag has equivalent dead time.

erg

erg (erg) is a non-SI (cgs) unit of energy, equal to 10^{-7} J.

error

error = setpoint - PV. In auto mode, the controller uses the error in its calculation to find the output that will get you to the setpoint.

esters

esters are organic compounds formed by reaction between alcohol and acids. Esters formed from carboxylic acids have the general formula RCOOR. Triesters, molecules containing three ester groups, occur in nature as oils and fats.

etch

a chemically caused change on the outside of a smooth floor surface which causes the floor to be pitted or rough.

etch allowance

the total dimensional adjustment, expressed as inches per side, incorporated into the artwork design for a photochemical machined part, compensating for the chem ical etching process.

etchant

the chemical etch involved with photo-etching a component.

etched blank

the pcm machined part in the flat or preformed configuration.

etching

chem. dissolution of material.

ethers

ethers are organic compounds with formula R-O-R, where R is not equal to H. They may be derived from alcohols by elimination of water, but the major method is catalytic hydration of olefins. They are volatile highly flammable compounds; when containing peroxides they can detonate on heating. The term ether is often used synonymously with ethyl ether.

ethoxylate

a substance that has had ethylene oxide added to render it more water soluble. There is some concern about the impact of the ethoxylation process on the environment because of the presence of 1,4 Dioxane, a by-product of the process and a suspected carcinogen.

■ Euler number

Euler number (*Eu*) is a dimensionless quantity used in fluid mechanics.

■ eutectic

eutectic is a solid solution consisting of two or more substances and having the lowest freezing point of any possible mixture of these components.
Eutectic point is the lowest temperature at which the eutectic mixture can exist in liquid phase. A liquid having the eutectic composition will freeze at a single temperature without change of composition.

■ eutectic phase

one of the two phases found in the eutectic structure.

■ eutrophication

an overgrowth of aquatic plants caused by an excess of nitrates, nitrites and phosphates. It results in a shortage of oxygen in the water, causing the death of aquatic life.

■ evaporation

a change of state from liquid to gaseous (vapor), due to the escape of molecules from the surface. A liquid which evaporates readily is described as volatile.

■ evaporation speed

expressed in relation to the evaporation rate of n-Butyl Acetate which is standardised at 1.0. All products with evaporation rates greater than 1.0 are faster evaporating than n-Butyl Acetate and conversely numbers lower than 1.0 indicate a slower rate.

■ evapotranspiration

the combined processes of evaporation and plant transpiration that remove water from the soil and return it to the air.

■ event

1. an occurrence that has a specific starting time and, usually, a duration shorter than the time being simulated in a model.
2. uncertain occurrences that take place within a short time relative to the time frame of the model.

■ event sequence

a series of actions and/or occurrences within the natural and engineered components of a geologic repository operations area that could potentially lead to exposure of individuals to radiation. An event sequence includes one or more initiating events and associated combinations of repository system component failures, including those produced by the action or inaction of operating personnel. Those event sequences that are expected to occur one or more times before permanent closure of the geologic repository operations area are referred to as Category 1 event sequences. Other event sequences that have at least one chance in 10,000 of occurring before permanent closure are referred to as Category 2 event sequences.

■ event tree

a structurally tree-like diagram that is useful in representing sequences of events and their possible outcomes. Each node, or branching point, represents an event, and each branch from that node represents one of its possible outcomes. Each possible pathway along the tree, from beginning to end of a given line of branching, represents a specific scenario.

■ event tree (analysis)

a graphical logical model that identifies possible outcomes following an initiating event. With suitable data it can be used to quantify the occurrence of an event.

■ excitory neurotransmitter

chemical that lowers threshold (smallest detectable stimulation) that will cause a cell to generate an impulse by making acceptance of positive ions easier.

■ exclusion limit

in SEC (Steric-exclusion chromatography), the upper limit of molecular weight (or size) beyond which molecules will elute at the same retention volume, called the exclusion volume. Many SEC packings are referred to by their exclusion limit. For example, a 10 column of porous silica gel will exclude any compounds with a molecular weight higher than 100,000, based on a polystyrene calibration standard.

■ exclusion volume (v_e)

the retention volume of a molecule on an SEC packing; all molecules larger than the size of the largest pore are totally excluded and elute at the interstitial volume of the column.

■ exothermic

a reaction is called exothermic if energy (heat) is released during the reaction.

■ exothermic reaction

the majority of chemical reactions are accompanied by the evolution of heat - these are known as exothermic reactions. The reaction between an acid and a base is exothermic, it is possible to feel the temperature change when one is added to the other.

■ Exr

A rapid exothermic reation results in an explosion, as occurs when TNT explodes.

■ expected behaviour

1. the nominal behaviour of the repository system and the geologic barrier in the absence of disruptive events.
2. the mean value of the probability distribution describing that behaviour.

■ expected value

a variable's mean, or average, outcome. The weighted average of the number of possible outcomes, with each outcome being weighted by its probability of occurrence. The mean of a probability distribution of a random variable that one would expect to find in a very large, random sample. The sum of the possible values, each weighted by its probability—the centre of the random variable's histogram (frequency distribution).

■ expectorant

promotes the thinning and ejection of mucus or exudate from the lungs, bronchi, and trachea; sometimes the meaning is extended to all remedies that quiet a cough. Examples: ginger, licorice, calamus, cardamom, cinnamon, cloves, sage, eucalyptus, thyme, wild cherry.

■ explosion

a release of energy sufficient to cause a pressure wave.

■ exposure (e)

the light quantity received per unit area, exposure expressed in intensity, spectral composition and duration terminology.

■ exposure pathway

the course a chemical or physical agent takes from the source to the exposed organism; describes a unique mechanism by which an individual or population can become exposed to chemical or physical agents at or originating from a release site.

■ extracolumn effects

the band-broadening effects of the chromatographic system outside of the column itself. Extracolumn effects must be minimised in order to maintain the efficiency of the column. Areas of band broadening can include the injector, connecting tubing, endfittings, frits, detector cell volume, and internal detector tubing. The variances of all of these contributions are additive.

■ extrinsic semiconductor

a semi-conducting material for which the electrical behaviour is determined by impurities.

■ extrusion

a forming technique whereby a material is forced, by compression, through a die orifice.

■ extrusive event

an igneous (volcanic) event occurring at the surface, i.e., a volcanic eruption. In repository performance analyses, molten material is assumed to intersect waste packages in the repository and cause a release of radionuclides. Compare *intrusive event.*

■ face-centreed cubic (fcc)

a crystal structure found in some of the common elemental metals. Within the cubic unit cell, atoms are located at all corner and face-centreed positions.

■ fading

lightening of the paint's colour, usually caused by exposure to light or heat.

■ Fahrenheit scale

Fahrenheit scale is the temperature scale in which 212 degrees is the boiling point of water and 32 degrees is the freezing point of water. The scale was invented in 1714 by a German physicist G.D. Fahrenheit (1686-1736).

32°F = 0°C
212°F = 100°C
1°F =(5/9)°C
$T(°C) = (5/9)[T(°F) - 32]$
$T(°F) = (9/5)T(°C) + 32$

■ farad

farad (F) is the SI derived unit of electric capacitance. The farad is the capacitance of an electric capacitor between the two plates of which there appears a difference of electric potential of one volt when it is charged by a quantity of electricity equal to one coulomb (F = C/V). The unit is named after the British scientist M. Faraday (1791-1867).

■ faraday constant

faraday constant (F) is the electric charge of 1 mol of singly charged positive ions.

$F = N_A \times e = 96487\ C\ mol^{-1}$

where N_A is Avogadro's constant ($6.022 \times 10^{23}\ mol^{-1}$) and e is the elementary charge (1.602×10^{-19} C).

■ farinaceous

having a mealy texture or surface.

■ fatty acids

fatty acids are aliphatic monocarboxylic acids characterised by a terminal carboxyl group (R-COOH). Natural fatty acids commonly have a chain of 4 to 28 carbons (usually unbranched and even-numbered), which may be saturated or unsaturated. The most important of saturated fatty acids are butyric (C4), lauric (C12), palmitic (C16), and stearic (C18). The most common unsaturated acids are oleic, linoleic, and linolenic (all C18).

The physical properties of fatty acids are determined by chain length, degree of unsaturation, and chain branching. Short-chain acids are pungent liquids, soluble in water. As chain length increases, melting points are raised and water-solubility decreases. Unsaturation and chain branching tend to lower melting points.

■ fault

a fracture or a fracture zone in crustal rocks along which there has been movement of the fracture's two sides relative to one another, so that what were once parts of one continuous rock stratum or vein are now separated.

■ fault displacement

movement along two sides of a fault relative to one another.

■ fault tree (analysis)

a method for representing the logical combinations of various system states which lead to a particular outcome (Top event). With suitable data it can be used to quantify the probability or frequency of an event.

■ feedback mechanisms

a mechanism that connects one aspect of a system to another. The connection can be either amplifying (positive feedback) or moderating (negative feedback).

■ fehm

a finite element heat and mass transport computer software code that is used for saturated zone flow and transport and unsaturated zone transport calculations.

■ fermi energy

for a metal, the energy corresponding to the highest filled electron state in the valence bond at 0 K.

■ fermi level

fermi level is the highest energy of occupied states in a solid at zero temperature. The Fermi level in conductors lies in the conduction band, in insulators it lies in the valence band, and in semiconductors it falls in the gap between the conduction band and the valence band. It is named after the Italian physicst Enrico Fermi (1901 - 1954).

■ fermion

fermion is any particle with spin equal to an odd multiple of 1/2. Fermion is a particle that obeys Fermi-Dirac statistics, e.g. electron, proton, neutron, muon, etc.

■ ferrites

ferrites are ceramic materials of nominal formula $MO.Fe_2O_3$, where M is a divalent metal (Co, Mn, NI, or Zn). The ferrites show either ferrimagnetism or ferromagnetism, but are not electrical conductors, and they are used in high-frequency circuits as magnetic cores, in rectifiers on memory and record tapes, and various related uses in radio, television, radar, computers, and automatic control systems.

■ ferroelectric

a dielectric material that may exhibit polarisation in the absence of an electric field.

■ ferromagnetism

ferromagnetism is a type of magnetism in which the magnetic moments of atoms in a solid are aligned within domains which can in turn be aligned with each other by a weak magnetic field. The total magnetic moment of a sample of the substance is the vector sum of the magnetic moments of the component domains. In an unmagnetised piece of ferromagnetic material the magnetic moments of the domains themselves are not aligned; when an external field is applied those domains that are aligned with the field increase in size at the expense of the others. Ferromagnetic materials can retain their magnetisation when the external field is removed, as long as the temperature is below a critical value, the Curie temperature. They are characterised by a large positive magnetic susceptibility.

■ fibre

any material that has been drawn into a cylinder with a length-to-diameter ratio greater than about ten.

■ Fick's law

Fick's law is the statement that the flux J of a diffusing substance is proportional to the concentration gradient, i.e.,

$$J = -D(dc/dx)$$

where D is called the diffusion coefficient.

■ filler

an inert foreign substance added to a matrix to improve or modify its properties.

■ **film formation**

the paint's ability to form a continuous dry film. This process is the result of the water or solvents evaporation and the coming together of the binder particles. A continuous dry film repels water.

■ **film sandwich**

a sandwich composed of photomasks made with flexible film materials.

■ **filtration**

the process of separating a solid from a liquid or gas by the use of a membrane that will allow only the liquid to pass through.

■ **(Final Boiling Point) FBP**

the highest temperature indicated on the thermometer inserted in the flask during a standard laboratory distillation. This is generally the temperature at which no more vapor can be driven over into the condensing apparatus.

■ **fire point**

the lowest temperature at which an oil vaporises rapidly enough to burn for at least 5 seconds after ignition, under standard conditions.

■ **firing**

a high-temperature heat treatment that increases the density and strength of a ceramic piece.

■ **fissile material**

material capable of undergoing fission with neutrons of any energy, including thermal, or slow, neutrons. The three primary materials in this category are uranium-233, uranium-235, and plutonium-239.

■ **fission**

the splitting of a nucleus into at least two other nuclei resulting in the release of two or three neutrons and a relatively large amount of energy.

■ **fission product**

any nuclide, either radioactive or stable, that arises from fission, including both the primary fission fragments and their radioactive decay products. Also called daughter product, decay product.

■ **fixative**

a material that helps to slow the rate of evaporation of components of perfume formulations, thereby making the fragrance last longer. Certain essential oils, for example, sandalwood, have fixative properties.

■ **fixed oil**

a fixed oil is chemically the same as a fat, but is generally liquid. Examples include almond oil, apricot kernel oil and grapeseed oil.

■ **flac**

a specialised computer code developed to solve soil and rock mechanics problems. It is used in conjunction with tough to form a thermal-hydrologic-mechanical code to analyse those coupled effects.

■ **flaking**

the detachment of pieces of paint from the substrate, caused by a loss of adhesion and elasticity. Also known as scaling.

■ flammable limit

the flammable limits refer to the conditions under which a mixture of a flammable material and air may catch fire or explode. If the percentage of flammable material in the air is between the minimum and maximum limits, the presence of a flame or a source of ignition is likely to lead to rapid combustion or explosion. Flammable limits for many materials are in the range 2 - 10%, but for some materials the limits are much wider. Ether (diethyl ether) for example, has flammable limits of 1.7 - 48%, which is an unusually wide range. This, coupled with the low boiling point of ether (34.6 °C) and high vapour pressure at room temperature (400 mm Hg at 18°C) means that it is easy to create a potentially explosive mixture of ether in air, and renders this compound an extreme fire hazard. The material for which flammable limits are quoted may be a flammable liquid, such as ether or methanol, or a powder or dust such as flour or grain dust. Flammable limits are also sometimes refered to as explosion limits.

■ flash point

the minimum temperature of a petroleum product or other combustible fluid at which vapour is produced at a rate sufficient to yield a combustible mixture. Specifically, it is the lowest sample temperature at which the air vapour mixture will 'flash' in the presence of a small flame. Flash point may be determined by following ASTM Methods.

Fire Point is the minimum sample temperature at which vapour is produced at a sufficient rate to maintain combustion. Specifically, it is the lowest sample temperature at which the ignited vapour persists in burning for at least 5 seconds. Since the fire point of commercial petroleum oils ordinarily run about 30 ° C above the corresponding flash point, they are omitted from petroleum product data. Flash and fire points have obvious safety connotations - the higher the test temperature, the less the hazard of fire or explosion. Of comparable significance is their value in providing a simple indication of volatility, where a lower flash point denotes a more volatile material. The dilution of a crankcase oil with a fuel, for example, lowers the flash point. Flash and fire points should not be confused with Auto-Ignition Temperature, the temperature at which combustion occurs spontaneously without an external source of ignition.

■ flash-drying

the quick evaporation of liquid from a granular or porous solid by a rapid reduction in pressure or an updraft of warm air.

■ fleck

a defect in the clear background for a phototool with a black pattern or within the clear pattern for a phototool with a black background.

■ flocculation

a reversible process in which a number of emulsion droplets stick together to form a cluster which can be broken up by mechanical action restoring the emulsion to its original form.

■ flow field

a fluid's distribution in and through an area, including its velocity and density, as a function of position and time.

■ flow rate (f)

the volumetric rate of flow of mobile phase through an LC column.

■ fluorescence

the slow emission of longer wavelength light following the absorption of shorter wavelength radiation. Fluorescence is common with aromatic compounds with several rings joined together. Phosphorescence is similar but may persist long after the stimulating radiation is extinguished.

■ fluoride

this inorganic substance and some of its compounds are used in toothpastes to slow tooth decay. In many countries it is added to the water supply prior to reaching the consumer. High doses can cause dental fluorosis, nausea, vomiting and even death.

■ fluorocarbons

a colourless, non-flammable gas or liquid that can produce mild upper respiratory tract irritation. Fluorocarbons are commonly used as a propellant in hairsprays.

■ flux

the rate of transfer of fluid, particles, or energy passing through a unit area per unit time.

■ flux corrections

often described as 'fudge factors' in climate modelling, flux corrections are numerical corrections to numerical climate model data used to keep the models from straying far from the actual data when past temperature data is input.

■ foam

foams are dispersions of gases in liquids or solids. The gas globule may be of any size, from colloidal to macroscopic, as in soap bubbles. Bakers' bread and sponge rubber are examples of solid foams. Typical liquid foams are those used in fire-fighting, shaving creams, etc. Foams made by mechanical incorporation of air are widely used in the food industry (e.g. whipped cream, egg white, ice cream, etc.). Foams can be stabilised by surfactants.

■ foam inhibitor

an additive which causes foam to dissipate more rapidly. It promotes the combination of small bubbles into large bubbles which burst more easily.

■ foot candle power

illumination power unit equal to luminous flux density of 1 lumen per sq. ft. striking surface.

force majeure

a standard clause which indemnifies either or both parties to a transaction whenever events reasonably beyond the control of either or both parties occur to prevent fulfilment of the terms of the contract.

forging

mechanical forming of a metal or alloy by heating and hammering.

formaldehyde

an antimicrobial used in many product groups. It is a toxic, colourless gas that is an irritant and a carcinogen. When combined with water, formaldehyde is used as a disinfectant, fixative or preservative. Formaldehyde is found in many cosmetic products and conventional nail care systems. Its use is banned or restricted in many countries.

four ball test

two test procedures based on the same principle:
Four-Ball EP Test (ASTM D2596)
Four-Ball Wear Test (ASTM D2266)
The three lower balls are clamped together to form a cradle upon which the fourth ball rotates in a vertical axis. The balls are immersed in the lubricant under investigation. The test is used to determine the relative wear-preventing properties of lubricants operating under boundary lubrication conditions. The test is carried out at a specified speed, temperature and load. At the end of the specified period, the average diameter of wear scar on the three balls is reported. The Four Ball EP Test is designed to evaluate performance under much higher unit loads. In this test, the top ball is rotated at a specified speed (1700+/-60 rpm), but temperature is not controlled. The loading is increased at specified intervals until the rotating ball seizes and welds to the other balls. At the end of each interval the average scar diameter is recorded. Two values are generally reported- Load Wear Index and Weld Point.

fractionation range

refers to the operating range of a gel or packing This is the range in which the packing can separate molecules based on their size. Molecules that are too large to diffuse into the pores are excluded. Molecules that can diffuse into all of the pores totally permeate the packing, eluting (unseparated) at the permeation volume.

fracture

a break in rock caused by mechanical stresses. A fracture along which there has been displacement of the sides relative to one another is called a fault. A fracture along which no appreciable movement has occurred is called a joint.

fracture continuum

a continuum that represents fluid flow and transport through numerous individual fractures by approximating them as continuous flow and transport fields.

fracture permeability

the capacity of a rock to transmit fluid through fractures in the rock.

fracture toughness

critical value of the stress intensity factor for which crack extensions occurs.

free energy

a thermodynamic quantity that is a function of both the internal energy and entropy of a system.

free radical

free radical is a molecular fragment having one or more unpaired electrons, usually short-lived and highly reactive. They can be produced by photolysis or pyrolysis in which a bond is broken without forming ions. In formulas, a free radical is conventionally indicated by a dot (Cl·). Free radicals are known to be formed by ionising radiation and thus play a part in deleterious degradation effects that occur in irradiated tissue. They also act as initiators or intermediates in oxidation, combustion, photolysis, and polymerisation.

freeze-drying

a process in which food or tissue is preserved, by drying it in a frozen state under high vacuum conditions; used in preparing medicines and laboratory specimens and in some forms of food preservation.

freezing point

see **melting point.**

frenkel defect

in an ionic solid, a cation-vacancy and cation-interstitial pair.

freon

the term widely used in commercial industry for fully halogenated and partially halogenated chlorofluorocarbons.

frequency

frequency (*ι*) is number of cycles of a periodic phenomenon divided by time. Hertz (Hz) is the SI derived unit, with a special name, for frequency, equal to s^{-1}. It is named after the German scientist Heinrich Hertz (1857-1894).

frequency distribution

data grouped into classes (or ranges of values within the overall set of values, such as 1 to 5, 5 to 10, 10 to 20, etc.), with the classes listed in a table (or other format) showing the number of data points that occur in each class.

frit

the porous element at either end of a column that serves to contain the column packing. It is placed at the very ends of the column tube or, more commonly, in the endfitting. Frits are made from stainless steel or other inen metal or plastic, such as porous PTFE or polypropylene.

frontal analysis

chromatographic technique that involves continuous addition of sample to the column. with the result that only the least-sorbed compound, which moves at the fastest rate. is obtained in a pure state. The second least-sorbed compound elutes with the first-eluting compound, the third least-sorbed compound with the first and second compound, etc., until the original sample is eluting at the column exit . Frontal analysis is seldom used and is mainly a preparative technique.

fronting

peak shape in which the front pan of a peak (before the apex) in a chromatogram tapers in advance of the remainder of the peak. There is an asymmetric distribution with a leading edge. The asymmetry factor for a fronting peak has a value < 1. The opposite effect is tailing. Fronting is related to the shape of the sorption isotherm.

fuel assembly

a number of fuel rods held together by plates and separated by spacers, used in a reactor. This assembly is sometimes called a fuel bundle.

fuel blending

the process of loading low heat output waste with high heat output waste in a waste package to balance its total heat output. This process applies only to commercial spent nuclear fuel.

fuel matrix

the physical form and composition of the substance that holds the fissile material.

fuel oil

the heavy distillates from the oil refining process; used as fuel for power stations, marine boilers.

fugacity

fugacity (f) is a thermodynamic function used in place of partial pressure in reactions involving real gases and mixtures. For a component of a mixture, it is defined by $d\imath = RT(\ln f)$
where $\imath$ is the chemical potential. The fugacity of a gas is equal to the pressure if the gas is ideal. The fugacity of a liquid or solid is the fugacity of the vapour with which it is in equilibrium. The ratio of the fugacity to the fugacity in some standard state is the activity.

fuller's earth

an absorptive clay consisting of fine siliceous material, used in face masks to draw out impurities from the skin. It may be used instead of talc.

fully halogenated chlorofluorocarbons

a halocarbon (a compound containing carbon and one or more halogens) fully saturated with halogens.

functional pattern

the phototooling configuration required in obtaining the designed part.

fungible

interchangeable. Products which can be commingled for purposes of pipeline shipment.

■ fungicide

a material that is able to kill at least some types of fungi.

■ fungistatic

a material inhibiting the growth and multiplication of some types of fungi.

■ gain (of the controller)

this is another way of expressing the 'P' part of the PID controller. GAIN = 100/(Proportional Band). The more gain a controller has the faster the loop responds and more oscillatory the process.

■ gain (of the process)

gain is defined as the change in input divided by the change in output. A process with high gain will react more to the controller output changing. For example, picture yourself taking a shower. You are the controller. If you turned the hot water valve up by half a turn and the temperature changed by 10 degrees this would be a higher gain process than if the temperature changed only 3 degrees.

■ gain margin

the difference in the logarithms of the amplitude ratios at the frequency where the combined phase angle is 180 degrees lag is the Gain Margin.

■ gal

gal is a non-SI unit of acceleration, equal to 1 cm/s^2. Also called galileo.

■ galvanic corrosion

electrochemical corrosion caused by the flow of electricity that occurs when two dissimilar metals with differing electrical potentials are near each other in the presence of a conductor such as water with solutes in it.

■ gamma radiation

gamma radiation is electromagnetic radiation of extremely short wavelength. Gamma radiation ranges in energy from about 10^{-15} to 10^{-10} J (10 keV to 10 MeV) (wavelength less than about 1 pm). Gamma rays are emitted by excited atomic nuclei during the process of passing to a lower excitation state.

Gamma rays are extremely penetrating and are absorbed by dense materials like lead and uranium. Exposure to gamma radiation may be lethal.

■ ganglia

a group of nerve cells forming a nerve centre, especially one located outside the brain or spinal cord.

■ gantry

a hoisting machine that slides along a fixed platform or track, either raised or at ground level.

■ gas

1. gas is a state of matter in which the matter concerned occupies the whole of its container irrespective of its quantity.

2. chromatography (also gas liquid chromatography, GLC) A method of analysis often used to

help determine the components of essential oils and other liquids.

3. a gas is a substance which takes the shape of its container and expands to completely fill it's container. There are several types of gases with slightly different behaviours. These are ideal gasses, real gasses, super critical fluids, plasmas and critical opalescent materials.

■ gasoil

designation for No.2 heating oils and diesel fuels. A clean distillate fuel oil.

■ gauge pressure

many pressure-measuring devices measure not absolute pressure but only difference in pressure. For example, a Bourdon-tube gauge indicates only the difference between the pressure in the fluid to which it is tapped and the pressure in the atmosphere. In this case, then, the reference pressure is actually the atmospheric pressure. This type of pressure reading is called *gauge pressure*. For example, if a pressure of 50 kPa is measured with a gauge referenced to the atmosphere and the atmospheric pressure is 100 kPa, then the pressure can be expressed as either

p = 50 kPa gauge or p = 150 kPa absolute.

■ gauss

gauss (G) is a non-SI unit of magnetic flux density (*B*). 1 G = 10^{-4} T.

■ Gaussian curve

a standard error curve, based on a mathematical function. that is a symmetrical, bellshaped band or peak. Most chromatographic theory assumes a Gaussian peak.

■ gaussian system of units

gaussian system of units is a hybrid system used in electromagnetic theory, which combines features of both the electrostatic cgs subsystem (esu) and electromagnetic cgs subsystem (emu). With three base units, it uses em units in magnetism and es units in electrostatics. This involves using the constant c (the velocity of light in vacuum) to interrelate these sets of units.

■ Gay-Lussac's law

see **Charles' law.**

■ gel

gels are colloids in which both dispersed and continuous phases have a three-dimensional network throughout the material, so that it forms a jelly-like mass (e.g. gelatine). One component may sometimes be removed to leave a rigid gel (e.g. heating silica gel).

■ Gel Filtration Chromatography (GFC)

size-exclusion chromatography (SEC) carried out with aqueous mobile phases. Generally refers to separations carried out on soft gels such as polydextrans. Most gel filtration separations involve biopolymers.

Gel Permeation Chromatography (GPC)

SEC carried out with organic mobile phases. Used for the separation and characterisation of polymers. SEC with aqueous mobile phases is referred to as aqueous GPC, or GFC.

gel point

the stage at which a liquid begins to take on the semisolid characteristics of a gel.

genii-s

a quasi-stochastic computer software code used to evaluate the dose from the migration of radionuclides in the biosphere that may affect humans through ingestion, inhalation, and direct radiation. It is used to develop biosphere dose conversion factors.

Gibbs free energy

gibbs free energy (G) is an important function in chemical thermodynamics, defined by

$$G = H - TS$$

where H is the enthalpy, S the entropy, and T the thermodynamic temperature. Gibbs free energy is the energy liberated or absorbed in a reversible process at constant pressure and constant temperature. Sometimes called Gibbs energy and, in older literature, simply 'free energy'.

Changes in Gibbs free energy, ÄG, are useful in indicating the conditions under which a chemical reaction will occur. If ÄG is negative the reaction will proceed spontaneously to equilibrium. In equilibrium position ÄG = 0.

Gibbs phase rule

gibbs phase rule is the relationship used to determine the number of state variables, usually chosen from among temperature, pressure, and species composition in each phase, which must be specified to fix the thermodynamic state of a system in equilibrium:

$$F = C - P + 2$$

where C is the number of components in a mixture, P is the number of phases, and F is the degrees of freedom, i.e., the number of intensive variables that can be changed independently without affecting the number of phases.

glass

an inorganic product of fusion which has cooled to a rigid condition without crystallising.

glass sandwich

a sandwich composed of photomasks made with rigid glass materials.

glass transition temperature

glass transition temperature is the temperature at which an amorphous polymer is transformed, in a reversible way, from a viscous or rubbery condition to a hard and relatively brittle one.

glass-ceramic

materials. A predominantly crystalline product created by the controlled crystallisation of glass; characterised by high thermal shock resistance and low thermal expansion.

glauber's salt

glauber's salt is sodium sulphate decahydrate ($Na_2SO_4.10H_2O$). Loses water of hydration at 100 °C. Energy storage capacity is more than seven times that of water.

Global Climate Model (GCM)

a computer program that, when amounts of various gases and various other factors are specified, gives a prediction of what the temperature should be.

glycerides

glycerides are esters of glycerol (propane-1,2,3-triol) with fatty acids, widely distributed in nature. They are by long-established custom subdivided into triglycerides, 1,2- or 1,3-diglycerides, and 1- or 2- monoglycerides, according to the number and positions of acyl groups.

glycols

glycols are compounds that have two hydroxyl groups present in each molecule. The simplest member is ethylene glycol

glycyrrhizic acid

an acid derived from liquorice root. It has antibacterial properties, which make it particularly suitable for use in deodorants.

gnaphalium leontopodium

edelweiss extract - has great antioxidant properties, helps protect against UV-radiation.

goldsim

a software code used as a probabilistic shell for the TSPA component models that, in combination, simulate potential long-term behaviour of the repository.

gradient elution

technique for decreasing separation time by increasing mobile phase strength over time during the chromatographic separation. Also known as solvent programming. Gradients can be continuous or stepwise. Binary, ternary, and quaternary solvent gradients have been used routinely.

graft copolymer

a copolymer that is a graft polymer.

graft macromolecule

a macromolecule with one or more species of block connected to the main chain as side-chains, these side-chains having constitutional or configurational features that differ from those in the main chain.

graft polymer

a polymer composed of graft macromolecules.

grain

grain (gr) is a non-SI unit of mass, equal to 64.79891 mg.

grains hardness

a measure of water hardness. The actual amount of dissolved calcium and magnesium salts measured in parts per million.

grashof number

grashof number (*Gr*) is a dimensionless quantity used in fluid mechanics, defined by

Gr = *l³gáÄTñ²/ç²*

where *T* is temperature, *ñ* is density, *l* is length, *ç* is viscosity, *á* is cubic expansion coefficient, and *g* is acceleration of gravity.

gravimetry

gravimetry is the quantitative measurement of an analyte by weighing a pure, solid form of the analyte. Since gravimetric analysis is an absolute measurement, it is the principal method for analysing and preparing primary standards. A typical experimental procedure to determine an unknown concentration of an analyte in solution is as follows:

- quantitatively precipitate the analyte from solution
- collect the precipitate by filtering and wash it to remove impurities
- dry the solid in an oven to remove solvent
- weigh the solid on an analytical balance
- calculate the analyte concentration in the original solution based on the weight of the precipitate.

gravitational constant

gravitational constant (G) is the universal constant in the equation for the gravitational force between two particles

$F = Gm_1m_2/r^2$

where *r* is the distance between the particles and m_1 and m_2 are their masses.

gray

gray (Gy) is the SI derived unit of absorbed dose of radiation. The gray is the absorbed dose when the energy per unit mass imparted to matter by ionising radiation is one joule per kilogram (Gy = J/kg). The unit is named after the British scientist Louis Harold Gray (1905-1965).

greenhouse gas (ghg)

a gas in the atmosphere that has the ability to absorb infrared radiation (that emitted by the Earth itself), causing the same kind of action as a blanket or the glass windows of a greenhouse; includes water vapour, nitrous oxide, methane, carbon dioxide, ozone, halogenated fluorocarbons, hydrofluorocarbons, and perfluorinated carbons

grignard reagents

grignard reagents are organomagnesium halides, RMgX, having a carbon- magnesium bond (or their equilibrium mixtures in solution with $R_2Mg + MgX_2$).

gross alpha activity

the total radioactivity due to alpha particle emissions from radionuclides estimated in the plume of contamination.

ground control

support of rock in the subsurface of the repository (e.g., rock bolts and steel sets).

ground support

the system (rock bolt with wire mesh, steel cast, etc.) used to line the main and emplacement drifts to minimise rock or earth falling into the drifts.

groundwater

water contained in pores or fractures in either the unsaturated or saturated zones below ground level.

groundwater flux

the volumetric rate of groundwater flow through a unit area of an aquifer.

guar hydroxypropyltrimonium chloride

a cationic surfactant derived from chemically modified guar gum.

guarana

a natural stimulant - provides energy.

guard column

a small column placed between the injector and the analytical column. Protects the analytical column against contamination by sample particulates and, perhaps. by strongly retained species. The guard column is usually packed with the same material as the analytical column and is often of the same i.d. It is much shorter, costs less, and is usually discarded when it becomes contaminated.

half-life

for a given reaction the half life $t_{1/2}$ of a reactant is the time required for its concentration to reach a value that is the arithmetic mean of its initial and final (equilibrium) value.

In nuclear chemistry, (radioactive) half life is defined, for a simple radioactive decay process, as the time required for the activity to decrease to half its value by that process.

$N = N_0/2$

Hall effect

the phenomenon whereby a force is brought to bear on a moving electron or hole by a magnetic field that is applied perpendicular to the direction of motion. The force direction is perpendicular to both the magnetic field and the particle motion directions.

halocarbon

halocarbon is a compound containing no elements other than carbon, one or more halogens, and sometimes hydrogen. The simplest are compounds such as tetrachloromethane (CCl_4), tetrabromomethane (CBr_4), etc. The lower members of the various homologous series are used as refrigerants, propellant gases, fire-extinguishing agents, and blowing agents for urethane foams. When polymerised, they yield plastics characterised by extreme chemical resistance, high electrical resistivity, and good heat resistance.

halogens

the halogens are the elements fluorine (F) chlorine (Cl), bromine (Br), iodine (I), and astatine (At). They are non-metals, and make up part of the 17 group in the periodic table. Compounds of these elements are called halogenides or halides. The halogens all have a strong unpleasant odor and will burn flesh. They do not dissolve well in water. The five elements are strongly electronegative. They are oxidising agents, with fluorine being the strongest and astatine being the weakest. They react with most metals and many non-metals. Halogens form molecules which consists of atoms covalently bonded. With increasing atomic weight there is a gradation in physical properties. For example: Fluorine is a pale green gas of low density. Chlorine is a greenish-yellow gas 1.892 times as dense as fluorine. Bromine is a deep reddish-brown liquid which is three times as dense as water. Iodine is a grayish-black crystalline solid with a metallic appearance. And astatine is a solid with properties which indicate that it is somewhat metallic in character.

■ hard water

water which contains calcium and magnesium salts that have dissolved from the rocks over which the water has flowed. Water that does not contain these salts is called soft water. There are two types of hardness temporary hardness, which can be removed relatively easily and permanent hardness, which is more difficult to remove.

■ hardenability

a measure of the depth to which a specific ferrous alloy may be hardened by the formation of martensite upon quenching from a temperature above the upper critical temperature.

■ hardness

hardness is the resistance of a material to deformation of an indenter of specific size and shape under known load. This definition applies to all types of hardness scales except Mohs scale, which is a based on the concept of scratch hardness and is used chiefly for minerals. The most generally used hardness scales are Brinell (for cast iron), Rockwell (for sheet metal and heat-treated steel), Knoop (for metals).

■ hardness of water

refers to water containing alkaline earth salts such as calcium and magnesium carbonate and bicarbonate. These prevent formation of a lather with soaps. Detergents can help remove the hardness of water, although water softening through ion exchange is a preferred method because it reduces the need for chemicals .

■ hawthorn juice

an excellent tonic for the heart and circulation.

■ hazard

a chemical or physical condition that has the potential for causing damage to people, property, or environment.

■ hazard and operability study

a systematic method for identifying possible hazards and potential operating problems in a plant or process by the application of so called 'guidewords' (e.g. more, less, other etc.) to the plant or process flow sheet to study process deviations.

■ hazardous chemical reactivity

any chemical reaction with the potential to exhibit rates of increase in temperature and/or pressure too high to be absorbed by the environment surrounding the system. Included are reactive materials and unstable materials.

■ head pressure

the pressure above gravity at the head of the column. Expressed in psi. bar. atm, or MPa

■ heart cutting

in preparative LC, refers to collection of the centre of the peak, where purity should be maximum. Also used in column switching.

■ heat

heat is a form of energy associated with and proportional to molecular motion. It can be transferred from one body to another by radiation, conduction, or convection.
Heat of atomisation is the energy required to dissociate one mole of a given substance into atoms.
Heat of combustion is the heat evolved when a definite quantity of a substance is completely oxidised (burned).
Heat of crystallisation is the heat evolved or absorbed when one mole of given substance crystallises from a saturated solution of the same substance.
Heat of formation is the heat evolved or absorbed when one mole of a compound is formed in their standard state from its constituent elements.
Heat of fusion is the heat required to convert a substance from the solid to the liquid state with no temperature change (also called latent heat of fusion or melting).
Heat of hydration is the heat evolved or absorbed when a hydrate of a compound is formed.
Heat of reaction is the heat evolved or absorbed as a result of the complete chemical reaction of molar amounts of the reactants.
Heat of sublimation is the energy required to convert one mole of a substance from the solid to the gas state (sublimation) without the appearance of the liquid state.
Heat of vaporisation is the heat required to convert a substance from the liquid to the gaseous state with no temperature change (also called latent heat of vaporisation).

■ heat capacity

heat capacity is defined in general as dQ/dT, where dQ is the amount of heat that must be added to a system to increase its temperature by a small amount dT. The heat capacity at constant pressure is $C_p = (DH/DT)_p$; that at constant volume is $C_V = (DE/DT)_V$, where H is enthalpy, E is internal energy, p is pressure, V

is volume, and *T* is temperature. An upper case *C* normally indicates the molar heat capacity, while a lower case *c* is used for the specific (per unit mass) heat capacity.

■ **heat of combustion gross**

total heat evolved during complete combustion of unit weight of a substance, usually expresses in BTU per pound.

■ **heat of combustion net**

gross heat of combustion minus the latent heat of condensation of any water produced.

■ **heat of reaction**

the total quantity of thermal energy liberated or absorbed during a chemical reaction.

■ **heat pipe**

a zone characterised by a continuous process of boiling, vapour transport, condensation, and migration of water back to the heat source.

■ **heat wait search (hws)**

an experimental technique in which a substance is heated in stages until very slow decomposition is detected. The experimental apparatus then becomes adiabatic and the course of the decomposition is followed.

■ **heavy crude**

crude oil with a high specific gravity and a low API gravity due to the presence of a high proportion of heavy hydrocarbon fractions and metallic content.

■ **heavy water**

a molecule of water contains two atoms of *hydrogen* and one of *oxygen*. Most water comprises of *hydrogen/oxygen* but a small percentage is composed of another *hydrogen* isotope, *deuterium* and *oxygen*. *Deuterium* differs from *hydrogen* by having one neutron in the nucleus of each atom. *Deuterium* water is given the formula $\mathbf{D_2O}$. There is approx. 1 part in 5000 $\mathbf{D_2O}$ in normal water and it can be concentrated by electrolysis. $\mathbf{D_2O}$ had a higher boiling point (101.4 °C) and melts at 3.6 °C

■ **hectare**

hectare (ha) is a unit of area equal to 10^4 m^2. The unit is still used in agriculture.

■ **hectorite**

a mineral found in bentonite clay used as a thickening agent and for its absorption and conditioning properties.

■ **Heisenberg uncertainty principle**

Heisenberg uncertainty principle is the principle that it is not possible to know with unlimited accuracy both the position and momentum of a particle. German physicist Werner Heisenberg (1901-1976) discovers this principle in 1927.

■ **Helmholz free energy**

Helmholz free energy (*A*) is a thermodynamic function defined by $A = U - TS$, where *U* is the internal energy, *S* the entropy, and *T* the thermodynamic temperature. For a reversible isothermal process ÄA represents the useful work available.

■ **haematopoietic agent**

a haematopoietic Agent is a chemical which interfers with the blood system by decreasing the oxygen-carrying ability of haemoglobin. This can lead to cyanosis and unconsiousness. Carbon monoxide is one such agent, familiar to smokers.

■ **Henry**

henry (H) is the SI derived unit of inductance equal to the inductance of a closed circuit in which an e.m.f. of one volt is produced when the electric current in the circuit varies uniformly at the rate of one ampere per second (H = V·s/A). The unit is named after the US physicst Joseph Henry (1797-1878).

■ **Henry's law**

Henry's law was discovered in 1801 by the British chemist William Henry (1775-1836). At a constant temperature the mass of gas dissolved in a liquid at equilibrium is proportional to the partial pressure of the gas. It applies only to gases that do not react with the solvent.

$x_i = K_x p_i$

where p_i is the partial pressure of component i above the solution, x_i is its mole fraction in the solution, and K_x is the Henry's law constant (a characteristic of the given gas and solvent, as well as the temperature).

■ **hepa filters**

this acronym stands for high efficiency particulate air filter.

■ **hertz**

hertz (Hz) is the SI derived unit, with a special name, for frequency. The hertz is the frequency of a periodic phenomenon of which the period is one second (Hz = 1 s^{-1}). It is named after the German scientist Heinrich Hertz (1857 - 1894).

■ **heterocyclic molecule**

a molecule containing at least one ring of *carbon* atoms, and which also has at least one atom from a different element, (*nitrogen*, *sulphur*, *oxygen* etc..) within the ring. Examples are shown below:

■ **h mol**

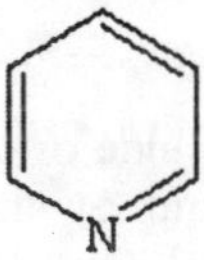

Pyridine

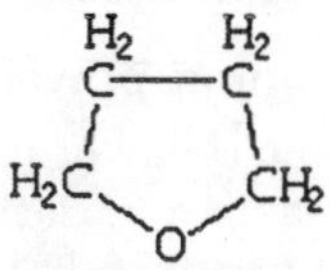

Tetrahydrofuran

■ **heterogeneous**

being composed of parts or elements of different kinds, such as a mixture of liquid-vapour or liquid-vapour-solid. A condition in which the value of a parameter, such as porosity of rock, varies over space and time.

■ **hetp**

height equivalent to a theoretical plate. A carryover from distillation theory, a measure of a column's efficiency. For a typical HPLC column well-packed with 5 μm particles, HETP (or *H*) values are usually between 0.01 and 0.03 mm. HETP = L/N, where L is column length and N is the number of theoretical plates.

heusler alloys

heusler alloys are alloys of manganese, copper, aluminum, nickel, and sometimes other metals which find important uses as permanent magnets.

Hexagonal Close-Packed (HCP)

a crystal structure found for some metals. The HCP unit cell is of hexagonal geometry and is generated by the stacking of close-packed planes of atoms.

HHV

higher heating value

hiding power

the ability of paint to hide or obscure a surface, colour or stain over which it has been uniformly applied. Hiding power is provided by the paint's pigment.

high-level radioactive waste

the highly radioactive material resulting from the reprocessing of spent nuclear fuel, including liquid waste produced directly in reprocessing; and any solid material derived from such liquid waste that contains fission products in sufficient concentrations; and other highly radioactive material.

high-level radioactive waste glass

the waste form of high-level radioactive waste in which the radioactive waste is mixed with borosilicate glass.

Hydrophile-Lipophile Balance HLB

a property of a surfactant which is represented by an arbitrary scale of 0-20 wherein the most hydrophilic materials have the highest numbers. The HLB of a nonionic surfactant is the approximate weight of ethylene oxide in the surfactant divided by 5.

hole (electron)

for semi-conductors and insulators, a vacant electron state in the valence band that behaves as a positive charge carrier in an electric field.

holidays

application defect whereby small areas are left uncoated.

holography

holography is a technique for creating a three-dimensional image of an object by recording the interference pattern between a light beam diffracted from the object and a reference beam. The image can be reconstructed from this pattern by a suitable optical system.

homogeneous

describes a substance which is the same throughout in its properties and composition.

■ **homogeniser**

a mechanical device which is used to create a stable, uniform dispersion of an insoluble phase (asphaltenes) within a liquid phase (fuel oil).

■ **homopolymer**

a polymer derived from one species of (real, implicit or hypothetical) monomer . Many polymers are made by the mutual reaction of complementary monomers. These monomers can readily be visualised as reacting to give an 'implicit monomer', the homopolymerisation of which would give the actual product, which can be regarded as a homopolymer. Common examples are poly (ethylene terephthalate) and poly (hexamethylene adipamide). Some polymers are obtained by the chemical modification of other polymers such that the structure of the macromolecules that constitute the resulting polymer can be thought of as having been formed by the homopolymerisation of a hypothetical monomer. These polymers can be regarded as homopolymers. Example: poly(vinyl alcohol).

■ **homopolymerisation**

polymerisation in which a homopolymer is formed.

■ **Hooke's law**

when a load is applied to any elastic body it is deformed or strained. Then the resulting stress (the tendency of the body to resume its normal condition) is proportional to the strain. Stress is measured in units of force per unit area, strain is the extent of the deformation

■ **hormones**

biological chemicals that act as 'messengers' in organisms by causing changes in biological processes

■ **horse power**

horse power is obsolete non-SI unit of power introduced by James Watt in 1782 to describe the power of steam machinery. It was equal to the work effort of a horse needed to raise vertically 528 cubic feet of water to one metre in one minute (HP = 735.498750 W).

■ **hot working**

any metal forming operation that is performed above a metal recrystallisation temperature.

■ **humalog**

a form of synthetic insulin that is absorbed faster than other synthetic forms

■ **human genome project**

a 15 year international undertaking consisting of research and technology development efforts designed to map and sequence some or all of the genome of humans.

■ **humectant**

a substance used to retain moisture on the skin and prevent the product from drying out.

■ **humidity**

a measure of moisture in the atmosphere. It depends on the

temperature and is higher in warm air than cold air.

humulus lupulus

hops, gentle skin toner.

hydration

hydration is addition of water or the elements of water (i.e. H and OH) to a molecular entity. The term is also used in a more restricted sense for the process:
A^+B^- à $A^+(aq) + B^-(aq)$

hydraulic conductivity

a measure of the ability to transmit water through a permeable medium. A number that describes the rate at which water can move through a permeable medium. The hydraulic conductivity depends on the size and arrangement of water-transmitting openings such as pores and fractures, the dynamic characteristics of the water such as density and viscosity, and the driving force.

hydraulic gradient

the difference in the height of water levels with respect to the distance between two locations.

hydraulic head

the pressure in the liquid expressed as equivalent height (head) of the water column.

hydraulic set cement

a cement that sets through reaction with water.

hydrocarbon

hydrocarbon is a compound containing only carbon and hydrogen.

hydrochloric acid

a strong acid used for pH adjustment. It is highly irritant in concentration. Also part of the stomach's digestive products.

hydrochlorofluorocarbon (HCFC)

a possible alternative to chlorofluorocarbons, consisting of the halogens chlorine and fluorine as well as hydrogen attached to a carbon backbone.
An alternative to CFCs; its direct global warming potential over 100 years is 1200-1600 times greater than that of carbon dioxide per volume.

hydrodynamic lubrication

a lubrication regime characterised by a full fluid film between two moving surfaces. The most common example is the type of lubrication which occurs in oil lubricated journal bearings. The movement of one surface (the shaft or journal) 'pulls' lubricating oil into the space between the journal and the bearing. This action causes a high pressure in the fluid which completely separates the two surfaces. By contrast, in boundary lubrication, there is only a partial fluid film separating the two surfaces and some surface-to-surface contact occurs.

hydrofluorocarbon (HFC)

an alternative to CFCs; its direct global warming potential over 100 years is 1200-1600 times greater than that of carbon dioxide per volume.

hydrogel

a water-based gel; a gel whose liquid constituent is water.

hydrogen bond

a strong secondary interatomic bond which exists between a bound hydrogen atom (its unscreened proton) and the electrons of adjacent atoms.

hydrogen ion

the ion which is left when the hydrogen atom loses its electron, forming a *proton*. In water the hydrogen ion binds to the water molecules, producing $\mathbf{H_3O^+}$. Compound which exhibit hydrogen bonding include water and amines.

hydrogen-induced cracking

occurs when atomic hydrogen generated at the surface of a metal migrates into the metal and forms hydride phases with the metal components, causing the metal to be more brittle. Hydride phases cause the metal to be more susceptible to cracking and, thus, to localised corrosion.

hydrologic properties

those properties of a rock that govern the capacity to hold, transmit, and deliver water, such as porosity, effective porosity, specific retention, permeability, and the directions of maximum and minimum permeabilities.

hydrolysis

hydrolysis is a chemical reaction in which water reacts with another substance to form two or more new substances. This involves ionisation of the water molecule as well as splitting of the compound hydrolysed, e.g. $CH_3COOC_2H_5 + H_2O$à $CH_3COOH + C_2H_5OH$
Examples are conversion of starch to glucose by water in presence of suitable catalysts and reaction of the ions of a dissolved salt to form various products, such as acids, complex ions, etc.

hydrometer

an instrument for determining the gravity of a liquid.

hydromethylcellulose

a chemically modified form of cellulose that is used as a thickening agent.

hydrophilic

'Water-loving' refers both to stationary phases that are compatible with water and to water soluble molecules in general. Most columns used to separate proteins are hydrophilic in nature and should not sorb or denature protein in the aqueous environment.

hydrophobic

refers both to stationary phases that are not compatible with water and to molecules in general that have little affinity for water. Hydrophobic molecules have few polar functional groups, most are hydrocarbons or have high hydrocarbon content.

hydrophobic interaction

hydrophobic interaction is the tendency of hydrocarbons (or of lipophilic hydrocarbon-like groups in solutes) to form intermolecular aggregates in an aqueous medium, and analogous intramolecular interactions. The name arises from the attribution of the phenomenon to the apparent repulsion between water and hydrocarbons. Use of the misleading alternative term hydrophobic bond is discouraged.

■ hydrophobic interaction chromatography

a technique in which reversed-phase packings are used to separate molecules by virtue of the interactions between their hydrophobic moieties and the hydrophobic sites on the surface. High salt concentrations are used in the mobile phase; separations are effected by changing the salt concentration. The technique is analogous to 'salting out' molecules from solution. Gradients are run by decreasing the salt concentration over time.

■ hydroscopic

capable of absorbing moisture from the atmosphere.

■ hydrosol

a colloidal dispersion in water; a sol whose dispersion medium is water.

■ hydrotreating

a generic name for a general refinery process for treating fuels or lubricant base stocks at elevated temperatures in the presence of hydrogen and a catalyst. Mild hydrotreating, sometimes called hydrofinishing, is used to improve the colour or odour of fuels and lubricating base stocks. Hydrotreating is a patented process which is used by a few manufacturers of superior lubricant base stock. In the process, the lubricant feedstock is reacted with hydrogen in the presence of a catalyst at a very high temperature (425° C) and high pressure (3200 psi). Under these severe conditions, virtually all olefin and aromatic hydrocarbons are cracked and saturated to yield a base stock which is 95-99+% saturated. Other impurities which contain sulphur, nitrogen, or oxygen are also destroyed by the severe hrdrocracking process. This Hydrotreating process produces very high quality ('semi-synthetic') lubricant base stock.

■ hydrotrope

a substance that increases the insolubility in water of another material, which is only partially soluble.

■ hygroscopic

describes a substance which can absorb up to 70% of its own mass of water vapour. Such a substance becomes damp, but does not dissolve.

■ hypertonic

a hypertonic solution contains a higher concentration of electrolytes than that found in body cells. If such a solution is allowed to enter the blood stream, the osmotic pressure difference between the blood and the cells will cause water to flow out of the cells, which will then shrink. This may cause serious harm, or even be fatal. Consequently, it is essential when blood transfusions are given, or blood replacement products are used, that the electrolyte concentration in the material to be given to a patient matches that of the body.

hypotonic

a hypotonic solution is one in which the concentration of electrolyte is below that in cells. In this situation osmotic pressure leads to the migration of water into the cells, in an attempt to equalise the electrolyte concentration inside and outside the cell walls. If the difference in concentration is significant, the cell walls may rupture, leading to the death of the cell. Consequently, it is vital that the electrolyte concentration of liquids used during blood transfusions be equal to that in cells.

hysteresis (magnetic)

the irreversible magnetic flux density-versus-magnetic field strength (B-versus-H) behaviour found for ferromagnetic and ferrimagnetic materials.

hysteresis (valve)

in a valve with loose linkages, the air signal to the valve will have to change by an amount equal to the hysteresis before the valve stem will move. Once the valve has begun to move in one direction it will continue to move if the air signal keeps moving in the same direction. When the air signal reverses direction, the valve will not move until the air signal has changed in the new direction by an amount equal to the hysteresis.

IBP

Initial Boiling Point. In a standard laboratory distillation, the temperature on the distillation thermometer at the moment the first drop of distillate falls from the condenser.

ideal gas

ideal gas is a gas in which there is complete absence of cohesive forces between the component molecules; the behaviour of such a gas can be predicted accurately by the ideal gas equation through all ranges of temperature and pressure. The concept is theoretical, since no actual gas meets the ideal requirement.

ideal gas law

the generalised ideal gas law is derived from a combination of the laws of Boyle and Charles. Ideal gas law is the equation of state

$$pV = RT$$

which defines an ideal gas, where p is pressure, V molar volume, T temperature, and R the molar gas constant (8.314 $JK^{-1}mol^{-1}$)

ideal solution

ideal solution is a solution in which solvent-solvent and solvent-solute interactions are identical, so that properties such as volume and enthalpy are exactly additive. Ideal solutions follow Raoult's law, which states that the vapour pressure p_i of component i is $p_i = x_i p_i^*$, where x_i is the mole fraction of component *i* and p_i^* the vapour pressure of the pure substance i.

■ illuminance (or illumination)

the result of luminous flux striking a surface. In English units, 1 lumen of flux falling on one square foot area is defined as one foot-candle illumination. In metric units, 1 lumen illuminating one square meter is the definition for 1 lux. This gives a direct conversion factor 10.76 lux = 1 foot-candle.

■ image

the functional pattern representation on a substrate:
a) drafting - as part of a master drawing or layout;
b) optical - as projected on a screen;
c) photographic - a photomask or the emulsion chemical on a film or plate;
d) photo resist film - exposed and developed coating on a substrate.

■ imidazolidinyl urea

an antimicrobial agent, commonly used as a preservative. It can release formaldehyde into formulations if not used correctly. It can cause sensitisation but is generally considered to be of low toxicity.

■ immiscible

two liquids are said to be immiscible, if when added together they do not mix but form two separate liquid phases.

■ impact energy

a measure of the energy absorbed during the fracture of a specimen of standard dimensions and geometry when subjected to very rapid (impact) loading. Charpy and Izod impact tests are used to measure this parameter, which is important in assessing the ductile-to-brittle transition behaviour of a material.

■ IMS

industrial methylated spirits, generally of ethanol, which has a chemical added to make it unsuitable for internal consumption.

■ in situ

in its natural position or place. The phrase distinguishes in-place experiments, conducted in the field or underground facility.

■ inclusions

undesired materials in a solid matrix.

■ incompressible

an incompressible fluid is one whose density is constant everywhere. All fluids behave incompressibly (to within 5%) when their maximum velocities are below Mach 0.3.

■ index of refraction

for a non-absorbing medium, index of refraction (n) is the ratio of the velocity of electromagnetic radiation *in vacuo* to the phase velocity of radiation of a specified frequency in the medium.

■ indicator

indicator is a substance used to show the presence of a chemical substance or ion by its colour. Acid-base indicators are compounds, such as phenolphtaleine and methyl orange, that change colour reversibly, depending on whether the solution is acidic or basic. Oxidation-reduction indicators are substances that show a reversible colour change between oxidised and reduced forms.

Indicator	Acid/Alkali Colour Change	pH range
Cresol red	red - yellow	0.2 - 1.8
Thymol blue	red - yellow	1.2 - 2.8
Bromophenol blue	yellow - blue	3.0 - 4.6
Methyl orange	red - orange	3.1 - 4.4
Congo red	blue - red	3.0 - 5.0
Methyl red	red - yellow	4.2 - 6.3
4-nitrophenol	colourless - yellow	5.6 - 7.6
Bromothymol blue	yellow - blue	6.0 - 7.6
Phenol red	yellow - red	6.8 - 8.4
Phenolphthalein	colurless - red	8.3 - 10.0
Brilliant blue	blue - yellow	10.8 - 12.0

■ induction period/time

time interval (starting at operating conditions) after which a runaway shows its maximum effects.

■ infil

a numerical software code used to analyse infiltration of precipitation.

■ infiltration

the process of water entering the soil at the ground surface and the ensuing movement downward. Infiltration becomes percolation when water has moved below the depth at which it can be removed (to return to the atmosphere) by evaporation or evapotranspiration.

■ infinite diameter effect

name given hy John Knox to the following phenomenon: At a certain column length, a sample injected into the centre of the packed bed spreads by radial diffusion but never reaches the column wall, where wall effects can cause band broadening. Knox showed that a sample peak collected in the exact centre of the column exit displayed a higher efficiency than a sample peak collected near the wall. The infinite diameter effect depends on column length, internal diameter, particle size, and mobile phase propenies.

■ infrared radiation

electromagnetic radiation whose wavelength is longer than that of visible light, and is responsible for the transmission of radiant heat.

■ infrared

electromagnetic energy usually defined as heat in the invisible light spectrum beyond the 7600A range.

■ Inverse Square Law

Illumination law which states that the surface illumination, via a point source is proportional to the source intensity and inversely proportional to the square distance from the source and the surface.

■ infusion

the liquid resulting from boiling plant material with water in order to extract water-soluble plant components.

■ **inherently safer**

a system is described as inherently safer if it remains in a non-hazardous situation after the occurrence of unintended deviations from normal operating conditions. Inherently safer is used, rather than inherently safe, because it is not possible to eliminate all hazards.

■ **inhibition**

a protective measure where the reaction can be stopped chemically by addition of another material.

■ **inhibitor**

an Inhibitor is a material which is added to a chemical to prevent an unwanted reaction. For example, 2,6-di-t-butyl-p-cresol may be added to tetrahydrofuran (THF) to prevent potentially dangerous polymerisation. Inhibitors are often added to chemicals which tend to undergo self-induced free-radical polymerisation.

■ **initiating event**

a natural or human-induced event that causes an event sequence.

■ **inlet**

the initial part of the column, where the solvent and sample enter. There is usually an inlet frit that holds the packing in place and, in some cases, protects the packed bed.

■ **in-line fillter**

a device that prevents particulate matter from damaging the column. Modern in-line filters can be placed between the injector and the column without contributing to band hroadening. A filter in this position is used to prevent sample particulates from entering the packed hed or the inlet frit.

■ **innage**

space occupied in a product container.

■ **instrinsic semiconductor**

a semiconductor material for which the electrical behaviour is characteristic of the pure material.

■ **insulator**

insulator is a material in which the highest occupied energy band (valence band) is completely filled with electrons, while the next higher band (conduction band) is empty. Solids with an energy gap of 5 eV or more are generally considered as insulators at room temperature. Their conductivity is less than 10^{-6} S/m and increases with temperature.

■ **integral action**

the 'I' part of the PID controller. With integral action, the controller output is proportional to the amount and duration of the error signal. If there is more integral action, the controller output will change more when error is present. If your units on integral are in 'time/rep' or 'time' then decreasing your integral setting will increase integral action. If your units on integral are in 'rep/time' or '1/time' then increasing your integral setting increases integral action.

■ **interfacial tension**

a measure of the molecular forces existing at the boundary between two phases. It is expressed in dynes/ cm. Liquids with low interfacial tension are more easily emulsified.

■ **intermetallic**

a compound of two metals that has a distinct chemical formula. The bonds in intermetallic compounds are often partly ionic.

■ **international system of units**

International System of Units (SI) is the unit system adopted by the General Conference on Weights and Measures in 1960 and recommended for use in all scientific and technical fields. It consists of seven base units (meter, kilogram, second, ampere, kelvin, mole, candela), plus derived units and prefixes.

■ **Interpenetrating Polymer Notwork (IPN)**

a polymer comprising two or more networks which are at least partially interlaced on a molecular scale but not covalently bonded to each other and cannot be separated unless chemical bonds are broken.

■ **interstitial volume**

the total volume of mobile phase within the length of the column. It is made up of the intraparticle volume (inside the packing itself) and interparticle volume (between the packing particles). Same as Void volume Also abbreviated V or V_m.

■ **intumescent coatings**

fire retardant coating which, when heated becomes plastic and produces nonflammable gasses, such as carbon dioxide and ammonia. The gasses are trapped by the film, converting it to a foam about fifty times as thick as the original paint film. At this stage, the film solidifies, resulting in a thick, highly insulating layer of carbon, which effectively protects the substrate from fire.

■ **invariant point**

a point on a binary phase diagram at which three phases are in equilibrium.

■ **inverted U-tube manometer**

inverted U-tube manometer is used for measuring pressure differences in liquids. The space above the liquid in the manometer is filled with air which can be admitted or expelled through the tap on the top, in order to adjust the level of the liquid in the manometer.

■ **inviscid**

not viscous.

■ **ion**

an electrically charged atom or group of atoms. If an atom or molecule gains an electron. It becomes a negatively charged *anion*, or if it loses an electron it becomes a positively charged *cation*.

■ **ion chromatography**

an ion-exchange technique in which low concentrations of anions or cations are determined using low capacity ion exchangers with weak buffers. Conductivity detectors are often used. Ion chromatography is practised in two forms. In suppressed IC, a second column is used to remove the buffer ions so that sample ions can be more easily detected, a membrane separator is sometimes used. In nonsuppressed IC, weakly conducting buffers at low concentration are carefully selected. and the entire emulent is passed through. The detector ions are detected above the background signal .

ion exchange

ion exchange is a process involving the adsorption of one or several ionic species accompanied by the simultaneous desorption (displacement) of one or more other ionic species.

ion exclusion

the process in which ionised solutes can be separated from unionised or partially ionised solutes using ion-exchange resins. Separation results from Donnan membrane potential, where ionic solutes exist at a higher concentration in solution than in the resin, whereas nonionic solutes are evenly distributed between the mobile phase and resin. Therefore. ionic solutes will move faster down the column than the nonionic solutes.

ion retardation

refers to the use of amphoteric ion-exchange resins that are used to retard ionic molecules and allow nonionic molecules or nonelectrolytes to elute preferentially.

ion suppression

buffering in an aqueous mobile phase at a particular pH to suppress solute ionisation. For example. one can suppress the ionisation of weak carhoxylic acids by adjusting the pH below their ionisation constant. Useful for improving peak shape of weak acids and bases in RPC.

ion-exchange capacity

the number of ionic sites on the packing that can take part in the exchange process. Exchange capacity is expressed in mequiv/g. Typical strong anion-exchange resin may have 3-5 mequiv/g capacity.

ion-exchange chromatography

a mode of chromatography in which ionic substances are separated on cationic or anionic sites of the packing. The sample ion (and usually a counterion) will exchange with ions already on the ionogenic group of the packing. Retention is based on the affinity of different ions for the site and on a number of other solution parameters (pH, ionic strength, counterion type, etc.).

ionic copolymerisation

a copolymerisation which is an ionic polymerisation.

ionic gel

gel that contains ionic groups attached to the colloid structure, preventing the groups from diffusing out into a surrounding medium.

■ **ionic polymerisation**

a chain polymerisation in which the kinetic-chain carriers are ions or ion-pairs.

■ **ionic strength**

ionic strength (*i* or *I*) is a measure of the total concentration of ions in a solution.

■ **ionisation energy**

ionisation energy is the minimum energy required to remove an electron from an isolated atom or molecule (in its vibrational ground state) in the gaseous phase.

■ **ionising radiation**

any radiation capable of displacing electrons from an atom or molecule, thereby producing ions.

■ **ionisation**

the process of adding electrons to, or removing electrons from, atoms or molecules, thereby creating ions. High temperatures, electrical discharges, and nuclear radiation can cause ionisation.

■ **ionising**

the process of creating an ion from a stable element or compound, either by adding or removing an electron or electrons. For chlorine, the process involves adding an electron to form a negatively charged chloride ion.

■ **ionomer**

a polymer composed of ionomer molecules.

■ **ionomer molecule**

a macromolecule in which a small but significant proportion of the constitutional units have ionisable or ionic groups, or both. Some protein molecules may be classified as ionomer molecules.

■ **ion-pair chomatography**

form of chromatography in which ions in solution can be paired or neutralised and separated as an ion pair on a reversed-phase column. Ion-pairing agents are usually ionic compounds that contain a hydrocarbon chain that imparts a certain hydrophobicity so that the ion pair can be retained on a reversed-phase column. Ion-pairing can also occur in normal-phase chromatography when one pan of the pair is loaded onto a sorbent. but this technique is not as popular as the RPC technique .

■ **ions**

an electrically charged particle, formed when an atom loses or gains one or more electrons to form a stable outer shell. All ions are either cations or anions

■ **irregular macromolecule**

a macromolecule , the structure of which essentially comprises the repetition of more than one type of constitutional unit , or a macromolecule the structure of which comprises constitutional units not all connected identically with respect to directional sense.

■ **irregular packing**

refers to the shape of a silica gel-based packing. Irregular silicas are available in micro-particulate sizes. The packings are made by grinding silica gel into small particles and then sizing them into narrow fractions using classification machinery. Spherical packings are now used more often than irregular packings in HPLC, but less-expensive irregular packings are still widely used in prep LC.

■ irregular polymer

a polymer composed of irregular macromolecules.

■ irreversible adsorption

when a compound that has a very strong affinity for the adsorbent is injected onto a column, it can be adsorbed so strongly that it cannot be eluted from the column. A chemical reaction between the sample and the surface of the adsorbent is an example of irreversible adsorption.

■ irreversible colloid

a pseudo-colloid with permanently attached radionuclides.

■ irritant

any compound that causes a negative skin reaction on application or sometimes after application.

■ irrotational

an irrotational fluid flow is one whose streamlines never loop back on themselves. Typically, only inviscid fluids can be irrotational. Of course, a uniform viscid fluid flow without boundaries is also irrotational, but this is a special (and boring!) case.

■ ISO

(International organisation for Standardisation) - An organisation which establishes internationally recognised standards for products, and test methods. One example is the ISO Viscosity Grade system for industrial oils.

■ isobar

isobar is a line connecting points of equal pressure on a graphical representation of a physical system.

■ isobars

isobars are nuclide having the same mass number but different atomic numbers. ^{54}Cr and ^{54}Fe, ^{112}Cd and ^{112}Sn are isobars.

■ isochore

isochore is a line or surface of constant volume on a graphical representation of a physical system.

■ isocratic

use of a constant-composition mobile phase in liquid chromatography.

■ isoelectric point

isoelectric point is the pH of a solution or dispersion at which the net charge on the macromolecules or colloidal particles is zero. In electrophoresis there is no motion of the particles in an electric field at the isoelectric point.

■ isoluitane

a propellant used in aerosol products.

■ isomerism

if a molecule possesses the same molecular formula, but the atoms are arranged in a different way, then they are called isomers. Isomers have different physical and chemical properties from each other. Optically active isomers differ only in the fact that one is the mirror image of the other, as shown with lactic acid below:

■ isomorphism

isomorphism is the existence of two or more substances that have the same crystal structure, so that they form solid solutions.

■ isoperibolic

a system in which the controlling jacket temperature is kept constant.

■ isoprene

the smallest building block of terpenes.

■ isotactic

a type of polymer chain configuration wherein all side groups are positioned on the same side of the chain molecule.

■ isotactic macromolecule

a tactic macromolecule, essentially comprising only one species of configurational base unit, which has chiral or prochiral atoms in the main chain in a unique arrangement with respect to its adjacent constitutional units. In an isotactic macromolecule, the configurational repeating unit is identical with the configurational base unit. An isotactic macromolecule consists of meso diads.

■ isotactic polymer

a polymer composed of isotactic macromolecules .

■ isotherm

isotherm is a line connecting points of equal temperature on a graphical representation of a physical system.

■ isothermal

a system condition in which the temperature remains constant. This implies that potential temperature increases and decreases are compensated by sufficient heat exchange with the environment of the system.

■ isothermal calorimeter

a calorimeter in which the energy exchange with its surroundings is measured whilst the temperature of its contents remains essentially constant.

■ isothermal process

isothermal process is a thermodynamic process in which the temperature of the system does not change.

■ isotones

isotones are nuclides having the same neutron number N but different atomic number Z.

■ isotope

one of two or more atomic nuclei with the same number of protons (i.e., the same atomic number) but with a different number of neutrons (i.e., a different atomic weight). For example, uranium-235 and uranium-238 are both isotopes of uranium.

■ isotropy

the condition wherein all significant physical properties are equal when measured in any direction or along any axes.

■ IUPAC

international union of pure and applied chemistry (IUPAC) is a voluntary nonprofit association of national organisations representing chemist in 45 member countries. It was formed in 1919 with the object of facilitating international agreement and uniform practice in both academic and industrial aspects of chemistry.

■ izod impact test

one of two tests that may be used to measure the impact energy of standard notched specimen.

■ j-13 water

groundwater taken from Well J-13 or water made in a laboratory that has the same chemical composition.

■ jacket temperature

the temperature of the fluid contained in the reactor jacket used to indirectly heat or cool the reactor contents.

■ joint

a fracture in rock, usually more or less vertical to bedding, along which no appreciable movement has occurred.

■ joint set

in a rock mass, a group of parallel joints (fractures without displacement).

■ joule

joule (J) is the SI derived unit of energy, work, and heat. The joule is the work done when the point of application of a force of one newton is displaced a distance of one metre in the direction of the force (J = N m). The unit is named after the British scientist James Prescott Joule (1818-1889).

■ Joule-Thomson coefficient

Joule-Thomson coefficient (*ì*) is a parameter which describes the temperature change when a gas expands adiabatically through a nozzle from a high pressure to a low pressure region.

■ junction unit

a non-repeating atom or non-repeating group of atoms between blocks in a block macromolecule.

■ k'

capacity factor; k' = *(mass in stationary phase) (mass in mobile phase)*. Can be calculated from the equation $k' = (t_R - t_0)/t_0$, where t_R is retention time for the sample peak, and t_0 is the retention time for an unretained peak.

■ kaolin

a white powdery clay, also known as china clay, which arises following decomposition of feldspars in granites. It is used in face powder and clay masks for its absorption properties.

■ kelvin

kelvin (K) is the SI base unit of thermodynamic temperature. The kelvin, unit of thermodynamic temperature, is the fraction 1/273.16 of the thermodynamic temperature of the triple point of water. The unit is named after the

British scientist Sir. W. Thompson, Lord Kelvin (1824-1907).

■ **keralolytic**

an agent that causes skin shedding and removes surface layers of skin.

■ **keratin**

a fibrous protein found in hair and nails.

■ **ketone**

an organic compound containing the carbonyl group, **>C=O**, to which other carbon atoms are attached. Ketones are extremely useful as solvents and paint components. An example is Butanedione.

■ **ketones**

ketones are compounds in which a carbonyl group is bonded to two carbon atoms: $R_1R_2C=O$ (neither R may be H). They are derived by oxidation of secondary alcohols. The simplest member of the series is acetone, $(CH_3)_2CO$.

■ **ketosis**

ketosis is an excess of ketones in the body. This can be brought about by exposure to certain types of chemicals.

■ **kieselguhr**

a diatomaceous earth used both in column chromatography and as a sample cleanup medium. Only weakly adsorptive, it is used as a support in liquid-liquid chromatography. Rarely used in HPLC.

■ **kilogram**

kilogram (Kg) is the SI base unit of mass; it is equal to the mass of the international prototype of the kilogram

■ **kinematic viscosity:**

the ratio of the absolute viscosity of a liquid to its specific gravity at the temperature at which the viscosity is measured. Expressed in Stokes or Centistokes.

■ **kinetic data**

data associated with the conversion rate of a reaction, such as the activation energy, pre-exponential factor and order of reaction.

■ **kinetics**

(classical kinetics) The study of the relationship between motion and the forces affecting motion.

■ **kjeldhal flask**

kjeldhal flask is a round bottom flask with a long wide neck that is used in the determination of nitrogen by Kjeldhal's method.

■ **kjeldhal's method**

kjeldhal's method is an analytical method for determination of nitrogen in certain organic compounds. The method was developed by the Danish chemist Johan Kjeldahl (1849-1900).

It involves addition of a small amount of anhydrous potassium sulphate to the test compound, followed by heating mixture with concentrated sulphuric acid, often with a catalyst such as copper sulphate. As a result ammonia is formed. After alkalysing the mixture with sodium hydroxyde, the ammonia is separated by distillation, collected in standard acid, and the nitrogen determined by back-titration.

■ Knox equation

a modification of the van Deemter equation developed by John Knox.

■ knudsen number

Knudsen number (*Kn*) is a dimensionless quantity used in fluid mechanics, defined by

$Kn = \ddot{e}/l$

where $\ddot{e}$ is mean free path and l is length.

■ Kohlrausch's law

if a salt is dissolved in water, the conductivity of the solution is the sum of the conductances of the anions and cations. The law, which depends on the independent migration of ions, was deduced experimentally by German chemist Friedrich Kohlrausch (1840-1910).

■ ladder chain

a double-strand chain consisting of an uninterrupted sequence of rings, with adjacent rings having two or more atoms in common. A ladder chain is a double-strand chain with adjacent constitutional units joined to each other through four atoms, two on one side and two on the other side of each constitutional unit.

■ ladder macromolecule

a double-strand macromolecule consisting of an uninterrupted sequence of rings, with adjacent rings having two or more atoms in common. A ladder macromolecule is a double-strand macromolecule with adjacent constitutional units joined to each other through four atoms, two on one side and two on the other side of each constitutional unit.

■ lag time

lag time is the amount of time, after the dead time, that the process variable takes to move 63.3% of its final value after a step change in valve position. Lag time is also called a capacity element or a first order process. Very few real processes are pure lag. Almost all real processes contain some dead time.

■ laminar flow

laminar flow is smooth, uniform, non-turbulent flow of a gas or liquid in parallel layers, with little mixing between layers. It is characterised by small values of the Reynolds number.

■ lamination

a parts series or etched blanks which are stacked and bonded with proper registration, forming a complete unit.

■ lateral reversal

mirror image of the pattern's geometric orientation.

■ laneth-16

a mixture of ethoxylates produced from lanolin alcohol used for emulsifying and solubilising. It can be a mild irritant.

lanolin

a preparation of cholesterol and its esters obtained from wool fat. The lanolin is washed off the wool of living sheep. In the past many people thought they were allergic to lanolin; they were actually allergic to the detergent that was used to extract the lanolin from the wool. In recent years only mild detergent is used. Anhydrous lanolin refers to lanolin that does not contain water. Lanolin is an excellent emollient and thickener. It is protective to the skin and well tolerated by people with sensitive skin.

lanolin alcohol

a mixture of alcohols obtained by hydrolysis of lanolin, used for its moisturising properties.

lanthanides

lanthanides (lanthanons, lanthanoids or rare-earth elements) are a series of fourteen elements in the periodic table, generally considered to range in proton number from cerium to lutetium inclusive. It was convenient to divide these elements into the cerium group or light earth: cerium (Ce), praseodymium (Pr), neodymium (Nd), promethium (Pm), samarium (Sm), europium (Eu); and the yttrium group and heavy earths: gadolinium (Gd), terbium (Tb), dysprosium (Dy), holmium (Ho), erbium (Er), thulium (Tm), ytterbium (Yb) i lutetium (Lu). The position of lanthanum is somewhat equivocal and, although not itself a lanthanides, it is often included with them for comparative purpose. The lanthanides are sometimes simply called the rare earths. Apart from unstable Pm the lanthanides are actually not rare. Cerium is the 26. most abundant of all elements, 5 times as abundant as Pb. All are silvery, very reactive metals.

laser

laser (light amplification by stimulated emission of radiation) is a light amplifier usually used to produce monochromatic coherent radiation in the infrared, visible, and ultraviolet regions of the electromagnetic spectrum.

latent heat

latent heat (L) is the quantity of heat absorbed or released when a substance changes its physical phase at constant temperature (e.g. from solid to liquid at the melting point or from liquid to gas at the boiling point).

latex

stable dispersion of a polymeric substance in an essentially aqueous medium. Fine dispersion of rubber or resin, natural or synthetic, in water; the synthetic is made by emulsion polymerisation. (Strictly speaking, after polymerisation a latex is a solid dispersed in water, and therefore is not an emulsion. Latex and emulsion are often used synonymously in the paint industry.)

latex paint

water-thinned paint made with synthetic binders such as polyvinyl acetate or acrylic resins. In contrast to oil-based paint, latex paint dries fast, flows smoothly, and cleans up easily with water. High-performance latex paints contain 100% acrylic resins.

latin hypercube sampling

a sampling technique that divides the cumulative distribution function into intervals of equal probability and then samples from each interval.

lattice

the regular geometrical arrangement of points in crystal space.

lattice constants

lattice constants are parameters specifying the dimensions of a unit cell in a crystal lattice, specifically the lengths of the cell edges and the angles between them.

lattice energy

lattice energy is the energy per ion pair required to separate completely the ions in a crystal lattice at a temperature of absolute zero.

lattice parameter

the combination of unit cell edge lengths and interaxial angles that defines the unit cell geometry.

lauramide mipa

a non-ionic surfactant used for emulsifying properties.

laureth 1 to 23

products of the combination of ethylene oxide with lauryl alcohol to enhance emulsifying properties.

Le Chatelier's principle

the idea that a system at equilibrium will respond to a stress placed upon it in such a manner as to partially offset that stress. The principle was first stated in 1888 by French physical chemist Henri Le Chatelier (1850-1936).

lever rule

mathematical expression whereby the relative phase amounts in a two-phase alloy at equilibrium my be computed.

lewis acid

lewis acid is an agent capable of accepoting a pair of electrons to form a coordinate bond.

lewis base

lewis base is an agent capable of donating a pair of electrons to form a coordinate bond.

LHV

lower heating value.

lifting

refers to tankers and barges taking on cargoes of oil or refined product at the terminal or transshipment point.

ligand

in ligand-exchange chromatography refers to the molecule added to the mobile phase that acts as the chelating agent In affinity chromatography, refers to the biospecific material (enzyme. antigen, or hormone) coupled to the suppon (carrier) to form the affinity column.

ligand field theory

ligand field theory is a description of the structure of crystals containing a transition metal ion surrounded by nonmetallic ions (ligands). It is based on construction of molecular orbitals involving the *d*-orbitals of the central metal ion and combinations of atomic orbitals of the ligands.

ligand-exchange chromatography

a technique in which chelating ligands are added to the mobile phase and undergo sorption onto a packing. These sorbed molecules can act as chelating agents with solutes. An example would be the use of copper amine chelates for the separation of amino acids Chelating resins function in a similar manner. Chelating groups are chemically bonded to the polystyrene backbone.

light year

light year (ly) is a unit of distance used in astronomy, defined as the distance light travels in one year in vacuum (ly = 9.46052973 x 10^{15}m)

light

electromagnetic energy defined as visible energy between 3800-7800 Ångstrom range.

lignin1. a polymer occurring with cellulose in lignified plant tissues.

2. lignins are macromolecular constituents of wood related to lignans, composed of phenolic propylbenzene skeletal units, linked at various sites and apparently randomly.

line loading

placing the waste packages at very close distances end-to-end to achieve a more uniform thermal profile along the length of the emplacement drifts.

linear chain

a chain with no branch points intermediate between the boundary units.

linear copolymer

a copolymer composed of linear macromolecules.

linear elution adsorption chromatography

a term coined by Lloyd Snyder. Refers to adsorption chromatography carried out in the linear portion of an adsorption isotherm.

linear macromolecule

a macromolecule , the structure of which essentially comprises the multiple repetition in linear sequence of units derived, actually or conceptually, from molecules of low relative molecular mass.

linear polymer

a polymer composed of linear macromolecules .

linear velocity (u)

the velocity of the mobile phase moving through the column. Expressed in cm/s. Related to flow rate by the cross-sectional area of the column. Sometimes expressed as v.

■ **lipids**

1. lipids are a loosely defined term for substances of biological origin that are soluble in nonpolar solvents. They consist of saponifiable lipids, such as glycerides (fats and oils) and phospholipids, as well as nonsaponifiable lipids, principally steroids, natural substances of a fat-like nature.

■ **liposomes**

artificial microscopic sacs or vesicles used for the introduction of various agents into cells in vitro. They can be manufactured from a variety of substances, including phospholipids. Some manufacturers claim they can be used to deliver certain substances to the underlying layers of the skin.

■ **liquid form of matter**

a liquid assumes the shape of its container. The molecules of a liquid are in constant motion and do not have the fixed arrangement found in solids.

■ **Liquid-Liquid Chromatography (LLC)**

same as Partition chromatography The earliest form of HPLC, it gave way to chemically bonded phases in the early 1970s.

■ **liquid-phase sintering**

the progressive fusion of a ceramic material during the firing process; as it proceeds the proportion of glassy bond increases and the porosity of the fired product decreases.

■ **Liquid-Solid Chromatography (LSC)**

same as adsorption chromatography.

■ **litre**

litre (l, L) is a synonym for cubic decimeter ($L = dm^3$).

■ **lithologic**

pertaining to rock features such as colour, texture, mineral content, and weathering characteristics.

■ **lithophysae**

small, bubble-like holes in the rock caused by volcanic gases trapped in the rock matrix as the ash-flow tuff cooled.

■ **lithophysal**

pertaining to tuff units with lithophysae, voids having concentric shells of finely crystalline alkali feldspar, quartz, and other materials that were formed by entrapped gas that later escaped.

■ **lithosphere**

lithosphere is the outer layer of the solid earth, extending from the base of the mantle to the surface of the crust.

■ **living copolymerisation**

a copolymerisation which is a living polymerisation .

■ **living polymerisation**

a chain polymerisation from which chain transfer and chain termination are absent.

■ **load upset**

an upset to the process (that is not from changing the set-point). A simple example- you are taking a shower and someone flushes the toilet. The temperature suddenly changes on you, the controller. Another example

you are injecting steam into flowing cold water to get lukewarm water, and the inlet cold water changes temperature.

■ loading

the amount of stationary phase coated or bonded onto a solid suppon. In liquid-liquid chromatography, the milligram amount of liquid phase per gram of packing. In BPC. the loading may be expressed in μmol/m^2 or in %C. See **Coverage**

■ localised corrosion

a type of corrosion induced by local variations in electrochemical potential on a microscale over small regions. Variations in electrochemical potential may be caused by localised irregularities in the structure and composition of usually protective passive films on metal surfaces and in the electrolyte composition of the solution that contacts the metal.

■ log normal distribution

a distribution of a random variable x such that the natural logarithm of x is normally distributed.

■ logic tree methodology

an analytical approach that involves sequencing the studies in an analysis and then addressing certain attributes in each study.

■ lone pair

a pair of electrons not involved in bond formation.

■ long chain

a chain of high relative molecular mass.

■ long ton

an avoirdupois weight measure equalling 2,240 pounds.

■ longitudinal diffusion

same as molecular diffusion term; B-term in van Deemter equation.

■ longitudinal dispersion

dispersion of a solute moving in groundwater in the same direction as the groundwater flow path.

■ lookup table

a multidimensional table containing columns of data representing relationships between parameters in the table. A lookup table is a convenient way to represent and implement functional relationships between parameters considered in the model.

■ loose end

a chain, only one point of which is attached to a network .

■ low-level radioactive waste

radioactive waste producing small quantities of ionising radiation and that is not classified as high-level radioactive waste, transuranic waste, or byproduct tailings containing uranium or thorium from processed ore. Usually generated by hospitals, research laboratories, and certain industries.

LPG

lPG (Liquid Petroleum Gas) is a mixture of short hydrocarbons with most of the volume being propane and butane. LPG is considered an alternative fuel that burns cleaner than gasoline.

lumen

lumen (lm) is the SI derived unit of luminous flux. The lumen is the luminous flux emitted in a solid angle of one steradian by a point source having a uniform intensity of one candela (1lm =1 cd·sr).

luminance (or brightness)

flux reflected or emitted from a surface. The measurement has English units (foot lamberts) and metric units (candelas/square meter).

luminous energy

measurement for flux flow rate. The units are in lumen seconds.

luminous flux

luminous flux (Ö) is the intensity of light from a source multiplied by the solid angle. The SI unit is lumen.

lux

lux (lx) is the SI derived unit of illuminance. The lux is the illuminance produced by a luminous flux of one lumen uniformly distributed over a surface of one square metre (lx = lm·m^{-2}).

Lyman series

Lyman series is the series of lines in the spectrum of the hydrogen atom which corresponds to transitions between the ground state (principal quantum number n = 1) and successive excited states.

Mach

Mach number is the relative velocity of a fluid compared to its sonic velocity. Mach numbers less than 1 correspond to subsonic velocities, and Mach numbers > 1 correspond to super-sonic velocities.

machining

process involved with removing, forming, shaping, turning, reducing, finishing, milling material (any type) by various processes (chemically mill, milling machine tool, or others).

macrocycle

a cyclic macromolecule or a macromolecular cyclic portion of a macromolecule. In the literature, the term macrocycle is sometimes used for molecules of low relative molecular mass that would not be considered macromolecules.

macromolecule

polymer molecule A molecule of high relative molecular mass, the structure of which essentially comprises the multiple repetition of units derived, actually or conceptually, from molecules of low relative molecular mass. In many cases, especially for synthetic polymers, a molecule can be regarded as having a high relative molecular mass if the addition or removal of one or a few of the units has a negligible effect on the molecular properties. This statement fails in the case of certain macromolecules for which the properties may be critically dependent on fine details of the molecular structure. If a part or the whole of the molecule has a high relative molecular mass and essentially comprises the multiple repetition of units derived, actually or conceptually, from molecules of low relative molecular mass, it may be described as either macromolecular or polymeric, or by polymer used adjectivally.

■ **macromonomer**

a polymer composed of macromonomer molecules.

■ **macromonomer molecule**

a macromolecule that has one end-group which enables it to act as a monomer molecule, contributing only a single monomeric unit to a chain of the final macromolecule.

■ **macromonomeric unit**

macromonomer unit The largest constitutional unit contributed by a single macromonomer molecule to the structure of a macromolecule.

■ **macroporous resin**

cross linked ion-exchange resins that have both micropores of molecular dimensions and macropores several hundred Å wide. These are highly porous resins with large internal surface areas accessible to large molecules.

■ **macroradical**

a macromolecule which is also a free radical.

■ **magnesium aluminium silicate**

a component of natural dyes that is used as a thickener and suspending agent.

■ **magnetic field strength**

the intensity of an externally applied magnetic field.

■ **magnetic flux density**

the magnetic field produced in a substance by an external magnetic field.

■ **magnetic induction**

see **magnetic flux density.**

■ **magnetic susceptibility**

the proportionality constant between the magnetisation M and the magnetic field strength H.

■ **magnetisation**

the total magnetic moment per unit volume of material. Also, a measure of the contribution to the magnetic flux by some material within an H field.

■ **main chain**

backbone That linear chain to which all other chains, long or short or both, may be regarded as being pendant.

■ **manometers**

a somewhat more complicated device for measuring fluid pressure, consists of a bent tube containing one or more liquid of different specific gravities. Such a device is known as manometer.

In using a manometer, generally a known pressure (which may be atmospheric) is applied to one end of the manometer tube and the unknown pressure to be determined is applied to the other end.

In some cases, however, the difference between pressure at ends of the manometer tube is desired rather than the actual pressure at the either end. A manometer to determine this differential pressure is known as differential pressure manometer.

■ manual mode

in manual mode, the user sets the output.

■ Marine Diesel Oil (MDO)

Marine Diesel Oil is a middle distillate fuel oil which can contain traces often percent (10%) or more residual fuel oil from transportation contamination and/or heavy fuel oil blending. The MDO does not require heated storage.

■ martensite

a metastable Fe-C composition consisting of supersaturated carbon in iron that is the product of a diffusionless (athermal) transformation from austenite.

■ mask

a selective barrier to the passage of radiation. Masks, metal on glass: An optical mask comprising a glass substrate, selectively covered by thin, opaque metal layers, a photomask type.

■ masking

to inhibit with a reagent more active in a specific reaction.

■ mass

mass (m) is the quantity of matter contained in a particle or body regardless of its location in the universe. Mass is constant, whereas weight is affected by the distance of a body from the centre of the Earth (or of other planets). The SI unit is kilogram.

According to the Einstein equation

$$E = mc^2$$

all forms of energy possess a mass equivalent.

■ mass fraction

mass fraction (w_A) is the ratio of the mass of substance A to the total mass of a mixture.

■ mass numbers

mass number (A) is a characteristic property of a specific isotope of an element, equal to the sum of the number of protons and neutrons in the nucleus.

■ mass spectrometry

mass spectrometry is an analytical technique in which ions are separated according to the mass/charge (m/e) ratio and detected by a suitable detector.

In a mass spectrometer a sample is ionised and the positive ions produced are accelerated into a high-vacuum region containing electric and magnetic fields. These fields deflect and focus the ions onto a detector. A mass spectrum is thus obtained consisting of a series of peaks of variable intensity to which *m/e* values can be assigned. Different molecules can be identified by their characteristic pattern of lines.

■ mass transfer

the process of solute movement into and out of the stationary phase or mobile phase. The C-term of the van Deemter equation is referred to as the mass transfer term. The faster the process of mass transfer, the better the efficiency of the column. In HPLC, mass transfer is the most important factor affecting column efficiency. It is increased by the use of small-particle packings, thin layers of stationary phase, low viscosity mobile phases, and high temperatures.

■ master drawing

the technical drawing for the etched blank whose documented dimensions include all compensations for the chem machining process.

■ matrix

the body constituent of a composite or two-phase alloy that completely surrounds the dispersed phase and gives the body its bulk form.

■ matrix diffusion

the process by which molecular or ionic solutes, such as radionuclides in groundwater, move from areas of higher concentration to areas of lower concentration. This movement is through the pore spaces of the rock material as opposed to movement through the fractures.

■ matrix permeability

the capacity of the matrix to transmit fluid.

■ matrix porosity

in the solid, but porous, portion of rock, the ratio of openings, or voids, to the total volume of the matrix expressed as a decimal fraction or as a percentage.

■ matter

any substance that has mass (weight) and occupies space. It exists in any of three forms including a solid, liquid or gas.

■ maturating

an agent that promotes the maturing or bringing to a head of boils, carbuncles etc.

■ maximum pressure after decomposition

the maximum pressure which is obtained in a closed vessel. This pressure depends on the adiabatic temperature rise and the specific gas production.

■ Maximum residue limit

the Maximum residue limit is a value estimated for pesticides or veterinary drug residues in food. It is based on the assumption that good agricultural practice has been employed for the use of pesticides in farming, or good practice in the use of veterinary drugs. It is assumed that the product has been used in an appropriate manner and suitable withdrawal periods have been permitted. Since residues of pesticides and drugs may be broken down in tissue into various metabolites, MRLs are expressed either in terms of the amount of the parent compound remaining, or a metabolite that is representative of the residue of toxicological concern in the food. Estimates are usually made by comparing the acceptable daily intake (ADI) with a calculation of the total intake of the residue based on the MRLs and food intake data of these commodities for which MRLs have been established.

■ **MCR**

maximum continuous rating

■ **MDO**

marine diesel oil

■ **mean**

for a statistical data set, the sum of the values divided by the number of items in the set. The arithmetic average.

■ **mean free path**

mean free path is the average distance a gas molecule travels between collisions.

■ **mean pore diameter**

the average pore diameter of the pore in a porous packing. The pore diameter is imponant in that it must allow free diffusion of solute molecules into and out of the pore so that the solute can interact with the stationary phase. In SEC. the packings have different pore diameters, and therefore molecules of different sizes can be separated. For a typical adsorbent such as silica gel. 60-Å and 100-Å pore diameters are most popular. For packings used for the separation of biomolecules, pore diameters $\geq$ 300 Å are used.

■ **measurement**

same as 'process variable.'

■ **median**

a value such that half of the observations are less than that value and half are greater than the value.

■ **medicated products**

products that contain active ingredients to produce a specific physiological action.

■ **megabyte**

a megabyte is technically 2^20, or 1048580, bytes. Often, though, it is 10^6 bytes for marketing purposes.

■ **Meissner effect**

Meissner effect is the complete exclusion of magnetic induction from the interior of a superconductor.

■ **melting point**

melting point (or freezing point) is the temperature at which the solid and liquid phases of a substance are in equilibrium at a specified pressure (normally taken to be atmospheric unless stated otherwise). A pure substance under standard condition of pressure has a single reproducible melting point. The terms melting point and freezing point are often used interchangeably, depending on whether the substance is being heated or cooled.

■ meniscus

meniscus is the concave curve of a liquid surface in a graduate or narrow tube. Caused by surface tension.

■ mentha piperita

peppermint oil - antimicrobial, good for treating mouth ulcers and bad breath.

■ mer

the group of atoms that constitutes a polymer chain repeat unit.

■ mercaptans

mercaptans are a traditional term abandoned by IUPAC, synonymous with thiols. This term is still widely used.

■ mesogenic monomer

a monomer which can impart the properties of liquid crystals to the polymers formed by its polymerisation

■ meson

particle made of a Quark / Antiquark pair.

■ metal

the electroposite elements and alloys based on these elements.

■ metal hardening

a thermal, mechanical, or chemical treatment used to increase the hardness of a metal.

■ metallic

a substance that has, among other things, a good ability to conduct electricity.

■ metalloid

metalloid (semimetal) is any of a class of chemical elements intermediate in properties between metals and nonmetals. The classification is not clear cut, but typical metalloids are boron (B), silicon (Si), germanium (Ge), arsenic (As), and tellurium (Te). They are electrical semiconductors and their oxides are amphoteric.

■ metals

metals are a material in which the highest occupied energy band (conduction band) is only partially filled with electrons.

Their physical properties generally include:

- they are good conductors of heat and electricity. The electrical conductivity of metals generally decreases with temperature.
- they are malleable and ductile in their solid state.
- they show metallic lustre.
- they are opaque.
- they have high density.
- they are solids (except mercury)
- they have crystal structure in which each atom is surrounded by eight to twelve near neighbours

their chemical properties generally are:

- they have one to four valence electrons.
- they have low ionisation potentials; they readily lose electrons.

- they are good reducing agents.
- they have hydroxides which are bases or amphoteric.
- they are electropositive.

Metallic characteristics of the elements decrease and non-metallic characteristics increase with the increase of valence electrons. Also metallic characteristics increase with the number of electron shells. Therefore, there is no sharp dividing line between the metals and the non-metals.

Of the 113 elements now known, only 17 show primarily non-metallic characteristics, 7 others are metalloids, and 89 may be classed as metals.

metastable

nonequilibrium state that may persist for a very long time.

methane

a gas given off by livestock, wet rice agriculture, solid waste, coal mining, and oil and gas production; its direct global warming potential over 100 years is 11 times greater than that of carbon dioxide per volume.

metre

metre (m) is the SI base unit of length.

The metre is the length of the path travelled by light in vacuum during a time interval of 1/299 792 458 s.

metric ton heavy metal

a metric ton is a unit of mass equal to 1,000 kg (2,205 lb). Heavy metals are those with atomic masses greater than 230. Examples include thorium, uranium, plutonium, and neptunium. The term usually pertains to heavy metals in spent nuclear fuel and high-level radioactive waste.

metric ton

a weight measure equal to 1,000 kilograms, 2,204.62 pounds, and 0.9842 long tons.

mg/l

milligrams per liter = ppm (parts per million) - expresses a measure of the concentration by weight of a substance per unit volume.

mho

mho is an archaic name for the SI unit siemens (reciprocal ohm).

mica

silicate compounds used in eye colour cosmetics to create a glittery effect.

micellar chromatography

the addition of micelles to the mobile phase to effect separations. The micelles act as displacing or panitioning agents and provide another parameter that can be used to change selectivity.

micelle

micelle is an electrically charged colloidal particle, usually organic in nature, composed of aggregates of large molecules, e.g., in soaps and surfactants. For aqueous solutions, the hydrophilic end of the molecule is on the surface of the micelle, while the hydrophobic end (often a hydrocarbon chain) points toward the centre.

■ **micelle**

a spherical grouping of detergent molecules in water. Oils and greases dissolve in the hydrophobic centre of the micelle.

■ **micro lc**

refers collectively to techniques in which a column of smaller-than-usual internal diameter (i.d.) is used for separation.

■ **microbially influenced corrosion**

corrosion of the waste package that is induced by the activity of microbes.

■ **microbore**

refers to columns with smaller-than usual internal diameters (<2 mm) used in HPLC.

■ **microcalorimetry**

isothermal techniques of high sensitivity in which heat fluxes from the converting sample material are measured very accurately. Differential microcalorimetry is performed if the heat fluxes from the sample are compared with those of a reference material.

■ **microelectromechanical systems**

the use of silicon micromachining technologies to create complex functional devices such as pumps, valves, motors, actuators and the like, on the micron scale, by the selective etching of silicon. Also used to describe devices formed on glass and plastics

■ **microelectronics**

the branch of electronics that deals with components of miniature size.

■ **microfabrication technologies**

the use of photolithography tools and various additive and subtractive processes to define patterns that permit the building up of complex structures of different materials into a functional device.

■ **microfluidics technologies**

the movement of very small volumes of liquids within very narrow channels (100 microns or less) usually on silicon chips, but also on glass and plastic substrates. The performance of various chemical engineering unit operations such as separation, concentration, heat (heating or cooling) and mass transfer (mixing) within very narrow channels (100 microns or less) usually on silicon chips, but also on glass and plastic substrates. The development of pumps, sample collectors, connectors and the like for the handling of fluids and the small sample volumes (nanolitre to picolitre scale) associated with microfluidics.

■ **micron**

micron (*i*) is an obsolete name for micrometre (μm).

■ **micronetwork**

a highly ramified macromolecule that contains cyclic structures and is of colloidal dimensions.

■ microparticulate

refers to small particles used as HPLC stationary phases. Generally, packings with a particle diameter < 10 llm that are totally porous are considered microparticles.

■ microporous resin

same as microreticular resin.

■ microreactors

the selective etching of silicon to produce interconnected discrete volume elements wherein various chemical process operations may be performed. For example, a bioreactor may be formed by the immobilisation of an enzyme onto microshperes that are entrained within a reactor volume. Perfusion of the enzyme's substrate through the reactor results in the formation of product.

■ microreticular resin

cross-linked synthetic ion exchange resins with pore openings that correspond to molecular sizes. Diffusion into the narrow pores can be impaired, and low exchange rates, as well as poor performance, can occur, especially for large molecules.

■ microsphere

spheres of carboxylate-modified latex of varied diameters between 280 to 640 nm (0.000011 to 0.000025 in.), used as colloid tracers in the C-Wells testing.

■ microstructure

the structural features of an alloy that are subject to observation under a microscope.

■ microwave sounding unit (msu)

satellite used to study global temperatures. Orbiting weather satellites to detect the temperature for the stratosphere (denoted channel S), the middle troposphere (channel MT), and the lower troposphere (channel LT)

■ microwell technologies

the selective etching of silicon to produce a very high density of extremely small wells within silicon chips.

■ middle distillate

term applied to hydrocarbons in the so-called 'middle range' of refinery distillation. Examples: heating oil, diesel fuels, and kerosene.

■ mildewcide

chemical agent in quality paint that retards mildew, a common problem in humid climates.

■ Miller indices

a set of three integers that designate crystallographic planes, as determined from reciprocals of fractional axial intercepts.

■ Miller-Bravis indices

a set of four integers that designate crystallographic planes in hexagonal crystals.

■ millimetre of mercury

millimeter of mercury (mmHg) is a non-SI unit of pressure, equal to 133.322 Pa. The name is generally considered interchangeable with torr.

■ milling

process involved typically with larger area removal of material (any type) by various processes (mill chemically, mill machine tool, or others).

■ millirem (mrem)

a millirem is one one-thousandth of a rem, which is the unit of equivalent dose. Equivalent dose is a measure of the effect that radiation has on humans. The equivalent dose takes into account the type of radiation and the absorbed dose.

■ minimum plate height

the minimum of the curve that results from a plot of H vs. u. This value represents the most theoretical plates that can be obtained for a certain column and mobile phase system. Usually occurs at very low flow rates.

■ miocene epoch

the fourth of the five geologic epochs of the Tertiary Period, extending from the end of the Oligocene Epoch to the beginning of the Pliocene Epoch.

■ misalignment

improper relative image positioning. Molecular dye imaging materials: A particular diazo material sensitometrically designed for phototool applications by the manufacturer.

■ miscibility

a term often used interchangeably with solubility. It is the ability of a liquid or gas to dissolve uniformly in another liquid or gas.

■ miscible

two liquids are miscible if they dissolve completely in each other whatever the proportions of the components.

■ mitigating measures

measures to reduce the consequences of a runaway to an acceptable level.

■ mitigation

1. avoiding an impact by not taking a certain action or parts of an action.
2. minimising impacts by limiting the degree or magnitude of the action and its implementation
3. rectifying an impact by repairing, rehabilitating, or restoring the affected environment.
4. reducing or eliminating an impact over time by preservation and maintenance operations during the life of the action.
5. compensating for an impact by replacing or providing substitute resources or environments.

■ mixed dislocation

a dislocation that has both edge and screw components.

■ mixed waste

waste containing both radioactive hazardous substances and nonradioactive hazardous substances, regardless of whether these types of substances are combined chemically or mixed together.

■ mixed-bed column

combination of two or more stationary phases in the same column. used most ohen in IEC and SEC. Advantage in IEC is the total removal of ionic compounds useful in SEC because a wider molecular-weight range can be accommodated by the column.

■ mixture

a blend of two or more elements and/or compounds which are not chemically combine. A mixture can usually be separated into its elements or compounds fairly easily by physical means.

■ mmi

man Machine Interface. Refers to the software that the process operator 'sees' the process with. An example MMI screen may show you a tank with levels and temperatures displayed with bar graphs and values. Valves and pumps are often shown and the operator can 'click' on a device to turn it on, off or make a setpoint change.

■ mobile phase

the solvent that moves the solute through the column.

■ mode

auto, manual, or remote. In auto mode the controller calculates the output based its calculation using the error signal (difference between setpoint and PV). In manual mode, the user sets the output. In remote, the controller is actually in auto but gets its setpoint from another controller.

■ model

1. a conceptual description and the associated mathematical representation of a system, subsystem, component, or condition that is used to predict changes from a baseline state as a function of internal and/or external stimuli and as a function of time and space. 2. a depiction of a system, phenomenon, or process including any hypotheses required to describe the system or explain the phenomenon or process.

■ model validation

a process used to establish confidence that a conceptual model represented in a mathematical model by software or by other analytical means adequately represents the phenomenon, process, or system under consideration.

■ moderator

material that contains nuclei that cause energetic neutrons to slow down. In general, the lighter the element, the better it works as a moderator. Hydrogen, the lightest element, is a very efficient moderator; since water contains hydrogen nuclei, it, too, is a very effective moderator.

■ modifier

additive that changes the character of the mobile phase. For example. in reversed phase, water is the weak solvent; methanol, the strong solvent, is sometimes called the modifier

■ modulus of elasticity

the ratio of stress to strain for a material under perfectly elastic deformation.

molality

molality or molal concentration (*b*; formerly *m*) is a concentration in which the amount of solute is stated in moles and the amount of solvent in kilograms.
The unit of molality is mol kg^{-1}.

molar

denoting that an extensive physical property is being expressed per amount of substance, usually per mole.

molar mass

molar mass is the mass of one mole of a substance.
$M = m/n$
It is normally expressed in units of g mol^{-1}, in which case its numerical value is identical with the relative molecular mass.

molar quantity

molar quantity is often convenient to express an extensive quantity (e.g., volume, enthalpy, heat capacity, etc.) as the actual value divided by amount of substance (number of moles). The resulting quantity is called molar volume, molar enthalpy, etc.

molar volume

molar volume is the volume occupied by substance per unit amount of substance. The volume of the gas at 0°C and 101 325 Pa is 22.4 dm^3mol^{-1}.

molarity

formerly name for concentration.

mole

mole (mol) is the SI base unit of amount of substance.
The mole is the amount of substance of a system which contains as many elementary entities as there are atoms in 0.012 kg of carbon 12.
When the mole is used, the elementary entities must be specified and may be atoms, molecules, ions, electrons, other particules, or specified groups of such particles.
In this definition, it is understood that the carbon 12 atoms are unbound, at rest and in their ground state.

mole fraction

mole fraction (x_A) is the ratio of the amount of substance (number of moles) of substance A to the total amount of substance in a mixture.

molecular diffusion

the transfer of mass in a fluid by random molecular motion.

molecular diffusion term

refers to the B-term (second term) of the van Deemter equation. Also called longitudinal or axial diffusion term. Imponant in band broadening only at very low flow rates below the minimum plate height where the diffusion of individual solutes can occur in a longitudinal (lengthwise) direction on the column.

molecular weight

see **relative molecular mass.**

molecular weight distribution

the distribution of molecular weight of molecules in a polymer sample. Distribution can be defined as weight average and number average.

molecule

molecule is the smallest part of a chemical compound that can take part in a chemical reaction. The atoms of a molecule are held together by chemical bonds.

momentum

momentum (p) is the product of mass and velocity ($p = m v$).

monoatomic lattice

a crystal structure made up of only one type of atom; for instance, sodium metal forms a crystal structure of only sodium atoms

monodisperse colloidal system

a colloidal system having particles of the same size, interaction, and shape.

monomer

monomer is a simple molecule which is capable of combining with a number of like or unlike molecules to form a polymer. Monomer is repeating structure unit within a polymer.

monomer molecule

a molecule which can undergo polymerisation, thereby contributing constitutional units to the essential structure of a macromolecule.

monomeric phase

refers to a bonded phase in which single molecules are bonded to a Support. For silica gel, monomeric phases are prepared by the reaction of an alkyl or aryl monochlorosilane. Polymeric phases are generally prepared from a di- or trichlorosilane reactant.

monomeric unit

monomer unit mer The largest constitutional unit contributed by a single monomer molecule to the structure of a macromolecule or oligomer molecule . The largest constitutional unit contributed by a single monomer molecule to the structure of a macromolecule or oligomer molecule may be described as either monomeric, or by monomer used adjectivally.

monosaccharides

monosaccharide is a charbohydrate that cannot be split into smaller units by the action of dilute acids. Monosaccharide is any of several simple sugars having the formula $C_6H_{12}O_6$; the best-known are glucose, fructose, and galactose. Monosaccharides combine to form more-complex sugars known as oligo- and polysaccharide.

montmorillonite

pink clay - Absorbing and cleansing can help draw out impurities.

mosfet

metal-oxide-silicon field effect transistor, an integrated circuit element.

■ motor gasoline

a complex mixture of relatively volatile hydrocarbons with or without small quantities of additives, that have been blended to form a fuel suitable for use in spark-ignition engines.

■ mountain scale

1. similar to far-field for processes that are related to the area of the geosphere and biosphere far enough away from the repository that, when numerically modeled, show that releases from the repository are represented as a homogeneous, single source term. The effects of individual, small-scale components such as individual waste packages are not modeled because they are considerably smaller than the scale of the model.

2. a scale of hundreds of meters, or even kilometers, as opposed to tens of meters.

■ MSDS

material safety data sheet - a document that provides pertinent information and a profile of a particular hazardous substance or mixture. An MSDS is normally developed by the manufacturer or formulator of the hazardous substance or mixture. The MSDS is required to be made available to employees and operators whenever there is the likelihood of the hazardous substance or mixture being introduced into the workplace. Some manufacturers prepare MSDS for products that are not considered to be hazardous to show that the product or substance is not hazardous.

■ MTSR

abbreviation for maximum temperature of the synthesis reaction. The maximum temperature reached following the occurrence of the desired (synthesis) chemical reaction under adiabatic conditions starting at the designed process temperature. For a batch reactor the MTSR is equivalent to the process temperature plus the adiabatic temperature rise for the desired reaction. For a semi-batch reactor the MTSR is equivalent to the process temperature plus the adiabatic temperature rise of the synthesis reaction for those reactants that have accumulated during dosing.

■ mucilaginous

herbs that have a soothing effect on inflamed mucous membranes.

■ mucous membranes

thin layers of tissue that line the respiratory and genito-urinary tracts. They produce mucus as a protective film. They tend to be more sensitive and permeable to all types of preparations than other areas of skin.

■ mud

draws impurities through the skin and removes excess surface oil, dirt and grime. It can also be used to cleanse the hair. Mud packs have long been used for cleansing and therapeutic purposes.

■ **multidimensionai chromatography**

the use of two or more columns or chromatographic techniques to effect a better separation. Useful for sample cleanup, increased resolution, and increased throughput. It can be used off-line by collecting fractions and reinjecting onto a second column or on-line by the use of a switching valve. Also called coupled-column chromatography, column switching, multicolumn chromatography, or box-car chromatography.

■ **multi-strand chain**

a chain that comprises constitutional units connected in such a way that adjacent constitutional units are joined to each other through more than four atoms, more than two on at least one side of each constitutional unit. A chain that comprises constitutional units joined to each other through *n* atoms on at least one side of each constitutional unit is termed an *n*-strand chain, e.g., three-strand chain. If an uncertainty exists in defining *n*, the highest possible number is selected. 1.47 multi-strand macromolecule A macromolecule that comprises constitutional units connected in such a way that adjacent constitutional units are joined to each other through more than four atoms, more than two on at least one side of each constitutional unit. A macromolecule that comprises constitutional units joined to each other through *n* atoms on at least one side of each constitutional unit is termed an *n*-strand macromolecule, e.g., threestrand macromolecule. If an ambiguity exists in defining *n*, the highest possible number is selected.

■ **n**

the number of theoretical plates. $N = 16(t_R/W_b)^2$, where t_R is retention time, and W_b is the base width of the peak. A measure of the efficiency of a column. Sometimes measured as $N = 5.54(t_R/W_{1/2})^2$ where $w_{1/2}$ is the peak width at half height.

■ **nano-**

any device measured on the scale of nanometers.

■ **naphtha**

a volatile, colourless product of petroleum distillation. Used primarily as paint solvent, cleaning fluid, and blendstock in gasoline production, to produce motor gasoline by blending with straight-run gasoline.

■ **naphthenes**

one of three basic hydrocarbon classifications found naturally in crude oil. Naphthenes are widely used as petrochemical feedstock. Examples are: cyclopentane; methyl-,ethyl, and propylcyclopentane.

■ **narcotic**

a narcotic is a chemical agent which depresses the CNS (Central Nervous System) and may lead to unconciousness, and - potentially - death through respiratory failure. Many chemicals have narcotic properties, and exposure to them may give an initial feeling of euphoria before the depression.

narrow-bore column

columns of <0.5-mm i.d. used in HPLC. At this time. few columns of this type are commercially available.

natural

no official definition exists, but this term generally refers to products that exist in nature and are not further processed.

natural barriers

the physical components of the geologic environment that individually and collectively act to limit the movement of water or radionuclides.

natural colouring

food and cosmetic dyes which are unprocessed. The range of colours is limited and they are more prone to fading than synthetic dyes. The accepted natural colourings are less likely to be toxic than synthetic dyes.

natural variability

the tendency of the climate to change on time scales of 50 and 100 years without any of the underlying climate system-controlling factors having changed.

n-doping

when Silicon is doped with atoms of a group V element.

near field

the area and conditions within the repository including the drifts and waste packages and the rock immediately surrounding the drifts. The region around the repository where the natural hydrogeologic system has been significantly impacted by the excavation of the repository and the emplacement of waste.

néel temperature

néel temperature (T_N) is the critical temperature above which an antiferromagnetic substance becomes paramagnetic. The phenomenon was discovered around 1930 by L.E.F. Néel (1904-).

negative acting resist

a resist film which is polymerised (hardened) by light, and which, after exposure and development, remains on the substrate surface involving those areas which were under the transparent photomask's parts.

negative

the photo tool tone which has translucent functional patterns, namely, the desired metal portion of the part is rendered translucent.

neoprene

neoprene is a synthetic ruber made by polimerising the compound 2-chlorobuta-1,2-diene. Neoprene is often used in place of natural rubber in applications requiring resistance to chemical attack.

nessler's reagent

nessler's reagent is a solution of mercury(II) iodide (HgI_2) in potassium iodide and potassium hydroxide named after Julius Nessler (1827-1905). It is used in testing for ammonia, with which it forms a brown colouration or precipitate.

net infiltration

the water that has infiltrated down from the soil zone or exposed rock surface to a depth below which it cannot be removed by evapotranspiration. Net infiltration is the total infiltration at the surface minus water lost to evapouration and plant transpiration.

network

a highly ramified macromolecule in which essentially each constitutional unit is connected to each other constitutional unit and to the macroscopic phase boundary by many permanent paths through the macromolecule, the number of such paths increasing with the average number of intervening bonds; the paths must on the average be coextensive with the macromolecule. Usually, and in all systems that exhibit rubber elasticity, the number of distinct paths is very high, but, in most cases, some constitutional units exist which are connected by a single path only. If the permanent paths through the structure of a network are all formed by covalent bonds, the term covalent network may be used. The term physical network may be used if the permanent paths through the structure of a network are not all formed by covalent bonds but, at least in part, by physical interactions, such that removal of the interactions leaves individual macromolecules or a macromolecule that is not a network.

neut number or neutralistion number

the specific quanity of reagent required to 'neutralise' the acidity or alkalinity of a lube sample. Either of these characteristics - acidity or alkalinity - may be exhibited by an unused oil, depending on its composition. In addition, certain additives impart acidity, while alkalinity may be derived from the presence of detergents or of basic material added to control oxidation. In service, the oil will show increasing acidity as the result of oxidation, and additive depletion. Though acidity is not, by itself, necessarily harmful, an increase in acidity may be indicative of oil deterioration, and the neut number is widely used to evaluate the condition of the oil in service. The most common measurement is Acid Number, the specific quanity of KOH (potassium hydroxide) required to counterbalance the acid characteristics. How high an acid number can be tolerated depends on the oil and the service conditions. Only broad experimentation with the individual situation can determine the value. Neut number is determined in accordance with the ASTM method D664 or D974. The former is a potentiometric method, the latter, colourmetric. Values for Total Acid, Strong Acid, Total Base, and Strong Base can be obtained. Strong acid numbers are considered to be related to inorganic acids, such as those derived from sulfur, while the difference between the total and strong acid numbers is attributed to weak (organic) acids. A total acid number (TAN) and a total base number (TBN) can exist simultaneously, both representing components too weak to completely neutralise the other. When results are reported simply as 'neut number' or 'acid number', a Total Acid Number (TAN) is implied.

neutral

a chemical state that is neither acid nor alkali. A pH of 7 is considered neutral.

neutral cleaner

a floor cleaner that has a pH that is compatible with the finish to be cleaned. Generally this means a pH of between 7-9. Higher pH floor cleaners can attack the floor finish and dull it.

neutralisation

neutralisation is the process in which an acid reacts with a base to form a salt and water.
$H^+ + OH^-$ à H_2O

neutralisation number

the number that expresses the weight in milligrams of an alkali needed to neutralise the acidic material in one gram of oil. The neutralisation number of an oil is an indication of its acidity.

neutrino

neutrino is a stable elementary particle in the lepton family. Neutrinos have zero (or at least near-zero) rest mass and spin 1/2.

neutron

neutron is an elementary particle on spin 1/2 and zero charge. The free neutron has a mean lifetime of 887 seconds. Neutrons and protons, which are collectively called nucleons, are the constituents of the nucleus.

neutron absorber

a material (such as boron or gadolinium) that absorbs neutrons. Used in nuclear reactors, transportation casks, and waste packages to control neutron activity.

neutron logging

the analysis of the water content of soil and rocks in a borehole by means of neutron bombardment and the measurement of the reflected radiation.

neutron number

neutron number (*N*) is a characteristic property of a specific isotope of an element, equal to the number of neutrons in the nucleus.

newton

newton (N) is the SI unit of force, being the force required to give a mass of one kilogram an acceleration of 1 $m\ s^{-2}$ ($N = kg\ m\ s^{-2}$). It is named after British scientist Sir Isaac Newton (1642-1727).

Newtonian fluids

fluids which obey the Newton's law of viscosity are called as Newtonian fluids.
All gases and most liquids which have simpler molecular formula and low molecular weight such as water, benzene, ethyl alcohol, CCl_4,

hexane and most solutions of simple molecules are Newtonian fluids.

■ **NH_3N**

ammonia nitrogen.

■ **niacin**

vitamin B3.

■ **nitriles**

an organic cyanide yielding a corresponding acid and ammonia upon hydrolysis.

■ **nitrosamines**

compounds formed by the combination of a secondary amine with an oxide of nitrogen. They are highly carcinogenic in even minute quantities (parts per million). They are found in food products such as cured bacon, beer and also in cosmetics. In formulating products, care must be taken to prevent the formation of nitrosamines by monitoring pH and using antioxidant compounds. Some raw materials by their nature have the potential to cause nitrosamine formation and these must be especially closely monitored when used in formulae.

■ **nitrous oxide**

a greenhouse gas given off mainly by current agricultural methods; its direct global warming potential over 100 years is 270 times greater than that of carbon dioxide per volume

■ **NMR**

see **nuclear magnetic resonance.**

■ **noble gas**

noble gas refers to any element of the group of six elements in group 18 of the periodic table. They are helium (He), neon (Ne), argon (Ar), krypton (Kr) , xenon (Xe), and radon (Rn). Unlike most elements, the noble gases are monoatomic. The atoms have stable configurations of electrons. Therefore, under normal conditions they do not form compounds with other elements.
They were generally called inert gases until about 1962 when xenon tetrafluoride, XeF4, was produced in the laboratory. This was the first report of a stable compound of a noble gas with another single element.

■ **node**

a place where a wavefunction has a value of zero.

■ **nominal scenario or nominal case**

the performance assessment case, or conceptual model, representing the expected conditions of the disposal system as perturbed only by the presence of the repository, in the absence of disruptive events.

■ **noncrystalline**

the solid state wherein there is no long-range atomic order. Sometimes used synonymously with the terms amorphous, glassy and vitreous.

■ **nonflammable**

a nonflammable material is one which cannot be ignited (although the term is sometimes also used to indicate a material which, while it can be ignited with difficulty, burns only very slowly).

■ **nonionic surfactant**

a surface active agent that contains neither positively or negatively charged functional groups. These surfactants have been found to be especially effective in removing oily soil.

■ **non-ionic surfactant**

a surface active agent that has no electrical charge on it. Examples include polysorbate 20, cetyl stearyl alcohol. They can render formulations much milder.

■ **non-metals**

non-metals are defined as elements that are not metals.

Their physical properties generally include:

- they are poor conductors.
- they are brittle, not ductile in their solid state.
- they show no metallic luster.
- they may be transparent or translucent.
- they have low density.
- they form molecules which consists of atoms covalently bonded; the noble gases are monoatomic.

Their chemical properties are generally:

- they usually have four to eight valence electrons.
- they have high electron affinities (except the noble gases)
- they are good oxidising agents (except the noble gases)
- they have hydroxides which are acidic (except the noble gases)
- they are electronegative.

■ **non-Newtonian fluids**

fluids which do not obey the Newton's law of viscosity are called non-Newtonian fluids.

Generally non-Newtonian fluids are complex mixtures: slurries, pastes, gels, polymer solutions etc.

■ **nonoxynol-10**

a non-ionic emulsifier that is a derivative of phenol. It is irritant in concentration.

■ **nonporous particle**

refers to a solid particle used as a Suppon for a porous-coated or bonded phase.

■ **non-uniform flow**

non-uniform flow occurs when velocity, pressure etc., change from point to point in the fluid.

■ **non-uniform polymer**

polydicperse polymer A polymer comprising molecules non-uniform with respect to relative molecular mass or constitution or both.

■ **norepinephrine**

a neurotransmitter; a biogenic amine secreted by the adrenal medulla; a catecholamine that is very similar to dopamine.

■ **normalisation**

a mathematical manipulation that scales a function or a set of data to a particular value.

In the particle-in-a-box problem, we normalise the probability of finding the particle inside the box (psi squared of x) to one, since there is a 100% probability of finding it somewhere in the box.

■ normal-phase chromatography

a mode of chromatography carried out with a polar stationary phase and a nonpolar mobile phase. Adsorption on silica gel using hexane as a mobile nhase would be a typical normal-phase system. Also refers to the use of polar bonded phases, such as CN or NH_2. Sometimes referred to as straight-phase chromatography.

■ nosocomial pathogen

a pathogen that is found in hospital setting

■ notch:

a void; an undesirable indent within the photographic patterns edge, i.e., a clear indentation IN a black pattern; a black indentation IN a clear pattern.

■ nuclear magnetic resonance

nuclear magnetic resonance (NMR) is a type of radio-frequency spectroscopy based on the magnetic field generated by the spinning of electrically charged atomic nuclei. This nuclear magnetic field is caused to interact with a very large (1 T - 5 T) magnetic field of the instrument magnet. NMR techniques have been applied to studies of electron densities and chemical bonding and has become a fundamental research tool for structure determinations in organic chemistry. A technique of examining the bonds and distances between the atoms of a molecule using radio waves.

■ nuclear reactor

nuclear reactor is an assembly of fissionable material (uranium-235 or plutonium-239) designed to produce a sustained and controllable chain reaction for the generation of electric power. The essential components of a nuclear reactor are:

1. the core, metal rods containing enough fissionable material to maintain a chain reaction at the necessary power level (as much as 50 t of uranium may be required).
2. a source of neutrons to initiate the reaction (such as a mixture of polonium and beryllium)
3. a moderator to reduce the energy of fast neutrons for more efficient fission (material such as graphite, beryllium, heavy water, and light water are used)
4. a coolant to remove the fission-generated heat (water, sodium, helium, and nitrogen may be used)
5. a control system such as rods of boron or cadmium that have high capture cross sections (to absorbs neutrons)
6. adequate shielding, remote-control equipment, and appropriate instrumentation are essential for personnel safety and efficient operation.

■ nucleation

the initial stage in a phase transformation. It is evidenced by the formation of small particles (nuclei) of the new phase, which are capable of growing.

■ **nucleic acids**

nucleic acids are macromolecules (relative molecular mass over 5 000 000), the major organic matter of the nuclei of biological cells, made up of nucleotide units, and hydrolysable into certain pyrimidine or purine bases (usually adenine, cytosine, guanine, thymine, uracil), D-ribose or 2-deoxy-D-ribose.

■ **nucleon**

nucleon is a collective term for the proton and neutron.

■ **nucleus**

the very first few atoms that organise into a crystal structure; the crystal grows by adding on to the pattern established by these first few atoms

■ **nuclide**

nuclide is a species of atoms in which each atom has identical atomic number *Z* and identical mass number *A*.

■ **nuft**

a nonisothermal unsaturated zone flow and transport software code used for simulation of three-dimensional flow of groundwater, heat, and contaminant transport. It is used for drift-scale thermal-hydrologic calculations.

■ **nuisance material**

a nuisance material is one which can cause transient irritation or discomfort, but which has no long-term or systemic effects.

■ **numerical model**

an approximate representation of a mathematical model that is constructed using a numerical description method, such as finite volumes, finite differences, or finite elements. A numerical model is typically represented by a series of program statements that are executed on a computer.

■ **nymphaea alba**

water lily extract - excellent for soothing and calming sensitive skin.

■ **oat juice**

restorative for nervous exhaustion, derived from the fresh green plant. Can safely be used if you are allergic to gluten

■ **occlusion**

occlusion is the trapping of hazardous material next to the skin. This keeps the material in contact with the skin for long periods of time, increasing the chance that dermal damage will occur.

■ **occlusive agent**

a substance that prevent compounds from leaving a surface. On the skin, oils can act as occlusive agents by preventing the evaporation of water.

■ **occupational dose**

the radiation dose received by an individual in the course of employment in which the individual's assigned duties involve exposure to radiation or to radioactive material. Occupational dose does not include, for example, dose received from background radiation, from any medical administration the individual has received, from voluntary participation in medical research programs, or as a member of the public.

■ octadecylsilane (ods = rp-18).

the most popular reversed phase in HPLC. Octadecylsilane phases are bonded to silica or polymeric packings. Both monomeric and polymeric phases are available.

■ octahedral position

the void space among closed-packed, hard sphere atoms or ions for which there are six nearest neighbours. An octahedron (double pyramid) is curcumscribed by lines constructed from centres of adjacent spheres.

■ octyl dimethyl paba

see **paba.**

■ Oersted

Oersted (Oe) is a non-SI unit of magnetic field (*H*), equal to 79.57747157 A/m. The unit is named after H.C. Oersted (1777-1851).

■ Ohm

Ohm (W) is the SI derived unit of electric resistance. The ohm is the electric resistance between two points of a conductor when a constant difference of potential of one volt, applied between these two points, produces in this conductor a current of one ampere, this conductor not being the source of electromotive force (W = V/A). The unit is named after the German physicist Georg Simon Ohm (1789-1854).

■ olefins

olefins are acyclic and cyclic hydrocarbons having one or more carbon-carbon double bonds, apart from the formal ones in aromatic compounds. The class olefins subsumes alkenes and cycloalkenes and the corresponding polyenes.

■ oligomer

oligomer is a substance consisting of molecules of intermediate relative molecular mass (molecular weight), the structure of which essentially comprises the multiple repetition of units derived, actually or conceptually, from molecules of low relative molecular mass. In contrast to a polymer, the properties of an oligomer can vary significantly with the removal of one or a few of its units.

■ oligomer molecule

a molecule of intermediate relative molecular mass, the structure of which essentially comprises a small plurality of units derived, actually or conceptually, from molecules of lower relative molecular mass. A molecule is regarded as having an intermediate relative molecular mass if it has properties which do vary significantly with the removal of one or a few of the units. If a part or the whole of the molecule has an intermediate relative molecular mass and essentially comprises a small plurality of units derived, actually or conceptually, from molecules of lower relative molecular mass, it may be described as oligomeric, or by oligomer used adjectivally.

■ oligomerisation

the process of converting a monomer or a mixture of monomers into an oligomer. An oligomerisation by chain reaction carried out in the presence of a large amount of chain-transfer agent, so that the end-groups are essentially fragments of the chain-transfer agent, is termed telomerisation.

■ one-off sensitivity analysis

a method used to examine the effects of each component model or parameter on overall system performance. These analyses are conducted by fixing one important parameter of a particular component model at either its expected value, the median (or 50th percentile value), or at a specified extreme value (the 5th or 95th percentile value). These analyses are used to display the effect of the change on other measures of system or subsystem performance, as well as the effect on the variance of the projected dose history. By fixing a particular parameter that significantly affected the spread of the overall system performance results, it was possible to directly examine the significance of that parameter on the performance outcome.

■ onsager relations

onsager relations are an important set of equations in the thermodynamics of irreversible processes. They express the symmetry between the transport coefficients describing reciprocal processes in systems with a linear dependence of flux (J_i) on driving forces (X_j). In Onsager's theory the coupling coefficients are equal, $L_{ij} = L_{ji}$. This are known as reciprocal relations. The theory was developed by the Norwegian chemist Lars Onsager (1903-1976) in 1931.

■ onset temperature

defined as the temperature at which the heat released by a reaction can no longer be completely removed from the reaction vessel, and consequently, results in a detectable temperature increase. The onset temperature depends on detection sensitivity, reaction kinetics, vessel size and on cooling, flow and agitation characteristics. Scaling of onset temperatures and application of 'rules of thumb' concerning onset temperatures should be used with extreme care.

■ opacity

by definition, 1/T where T is transparency.

■ open loop

controller in manual mode.

■ open-tubular columns

columns of small internal diameter, currently being investigated for HPLC. SFC, and capillary electrophoresis. Stationary phases can be bonded on the internal walls of these columns. The most common type is the fused-silica tubing made for capillary GC.

■ optical lithography

using light (usually laser light) to alter a surface either chemically or physically

■ optically active resin

incorporation of optically active groups into an ion-exchange resin to allow separation of optically active isomers. Not many are commercially available in HPLC.

■ orbital

orbital is the area in space about an atom or molecule in which the probability of finding an electron is greatest.
The possible atomic orbitals correspond to subshells of the atom. Thus there is one *s*-orbital for each shell (orbital quantum number $l = 0$). There are three *p*-orbitals (corresponding to the three values of *l*) and five *d*-orbitals. The shapes of orbitals depend on the value of *l*.

■ order of magnitude

a range of numbers extending from some value to 10 times that value.

■ organic

describes products that have been certified as grown without the use of artificial pesticides and fertilisrs by the Soil Association or other certifying authority.

■ organic modifier

water-miscible organic solvent added to an aqueous mobile phase to effect separations in reversed-phase HPLC.

■ orientation

definition: The manner in which the functional pattern is to be viewed. Proper orientation requires 'ight reading'definition. The orientation for the various phototooling elements is described as either 'ight reading up'or 'ight reading down.'

■ orthochromatic

the emulsion silver component spectrally photo sensitive to blue, green, and yellow frequencies.

■ osmosis

osmosis is the flow of a solvent in a system in which two solutions of different concentration are separated by a semipermeable membrane which cannot pass solute molecules. The solvent will flow from the side of lower concentration to that of higher concentration, thus tending to equalise the concentrations. The pressure that must be applied to the more concentrated side to stop the flow is called the osmotic pressure.

■ osmotic pressure

osmotic pressure (*∂*) is the excess pressure necessary to maintain osmotic equilibrium between a solution and the pure solvent separated by a membrane permeable only to the solvent. In an ideal dilute solution

$\partial = c_B RT$

where c_B is the amount-of-substance concentration of the solute, R is the molar gas constant, and T the temperature.

■ ostwald's dilution law

ostwald's dilution law is a relation for the concentration dependence of the molar conductivity $\ddot{E}$ of an electrolyte solution.

The law was first put forward by the German chemist Wilhelm Ostwald (1853-1932).

■ outage

space left in a product container to allow for expansion during the temperature changes it may undergo during shipment and application. Measurement of space that is not occupied in a drum.

■ output

output of the PID controller. In auto mode the controller calculates the output based its calculation using the error signal (difference between setpoint and PV). In manual mode, the user sets the output.

■ overlay

a film containing graphics material which is used for inspection by superimposing the film on the graphics material.

■ overload

in preparative chromatography, the overload condition is defined as the mass of sample injected onto the column at which efficiency and resolution begin to be affected if the sample size is further increased.

■ overpotential

overpotential (ς) is a potential that must be applied in an electrolytic cell in addition to the theoretical potential required to liberate a given substance at an electrode. The value depends on the electrode material and on the current density.

■ oxidation

the term oxidation originally meant a reaction in which oxygen combines chemically with another substance. More generally, oxidation is a part of a chemical reaction in which a reactant loses electrons (increase oxidation number). Simultaneous reduction of a different reactant must occur (redox reaction).

■ oxidation inhibitor

chemical added in small quantities to a petroleum product to increase oxidation resistance, and lengthen its service or storage life. An oxidation inhibitor may combine with the peroxides and therefore modify the peroxides in such a way to arrest their oxidation influence. or the inhibitor (a passivator) may react with a catalyst either to 'poison' it or coat it with an inert film.

■ oxidising agent

an oxidising agent may be defined in various ways, depending upon the context in which the phrase is used. In broad terms it is often taken to mean a chemical which can act as an electron acceptor. Thus, in the reaction

$Zn^{2+}(aq) + 2Fe^{2+}(aq) = Zn(s) + 2Fe^{3+}(aq)$
the zinc is being reduced (gaining electrons) by reaction with the iron cations; the zinc in this instance is acting as an oxidising agent. This defintion is far more general than the one which follows if we regard 'oxidation' as being just the addition of oxygen to a substance. A variety of substances can act as oxidising agents. Oxygen on its own is a reasonably strong oxidising agent, but other materials, such as fluorine, metal nitrates, potassium permanganate, hydrogen peroxide, sodium hypochlorite (bleach), or sodium dichromate are very effective. Strong oxidising agents are often very reactive chemicals, and, in contact with combustible material such as paper, sawdust, fabrics or powdered metals, may form unstable mixtures which constitute a risk of fire or explosion.

■ oximes

oximes are organic compounds of structure $R_2C{=}NOH$ derived from condensation of aldehydes or ketones with hydroxylamine. Oximes from aldehydes may be called aldoximes; those from ketones may be called ketoximes.

■ oxo compounds

oxo compounds are organic compounds that contain the karbonyl group, C=O. The term thus embraces aldehydes, carboxylic acids, ketones, amides, and esters.

■ oxygen deficient atmosphere

an oxygen deficient atmosphere is one in which the level of oxygen is below that of normal air, around 19.5%.

■ ozokerite

a naturally occurring mineral wax that is used as a thickener. It has no known toxicity.

■ ozonation

the application of ozone to water, wastewater, or air, generally for the purposes of disinfection or odor control.

■ ozone

ozone is an alotropic form of oxygen. It is unstable blue gas with pungent odor. It is a powerful oxidising agent. The gas is made by passing oxygen through a silent electric discharge.
$3O_2(g)$à$2O_3(g)$

■ ozone layer

a region in the stratosphere where the mixing volume of ozone (O_3) is high. The layers of the atmosphere are distinguished by temperature. The stratosphere has high temperatures because O_3 is absorbing UV light, which is how the layer shields earth from UV radiation.

■ ozone-depletion potential

a measure of a chemical's destructiveness toward the ozone layer.

■ paba

a water-soluble acid (para-aminobenzoic acid) found in B vitamins and used as a sunscreen agent. Natural PABA is a sensitiser; synthetically produced octyl dimethyl PABA is not.

- **packing**

the adsorbent, gel, or solid Suppon used in the HPLC column. Most HPLC packings are <10 ym in average diameter.

- **pah**

polycyclic aromatic hydrocarbons. (occasionally polyaluminum hydroxide).

- **paired-ion chromatography**

same as lon-pair chromatography.

- **panchromatic**

the emulsion silver component spectrally photo sensitive to the entire visual frequency spectrum portion (red, green, and blue).

- **pantothenic acid (b5)**

a B-complex water-soluble vitamin, used in hair and skin-care products for its moisturising actions. It has no known toxicity.

- **paraffin wax**

wax derived from petrochemicals, used as a moisturiser and thickener.

- **paraffins**

paraffins are obsolescent term for alkanes. Still widely used in the petrochemical industry.

- **paramagnetism**

paramagnetism is a type of magnetism characterised by a positive magnetic susceptibility, so that the material becomes weakly magnetised in the direction of an external field. The magnetisation disappears when the field in removed.

- **parasiticide**

destroys parasites.

- **parrafin**

see **alkanes**.

- **partially halogenated chlorofluorocarbons**

a halocarbon only partially saturated with halogens. Ex.: hydrochlorofluorocarbons (which contain hydrogen as well as halogens).

- **particle size (dp)**

the average particle size of the packing in an LC column. A 5-mym column would be packed with particles having definite particle size distribution; packings are never monodisperse.

- **particle-size distribution**

a measure of the distribution of the particles used to pack the LC column. In HPLC, a narrow particle-size distribution is desirable. A particle-size distribution of dp +/- 10% would mean that 90% of the particles fall between 9 and 11 μm for an average 10-μm *dp* packing.

- **particulate**

free suspended solids.

- **particulates**

tiny pieces of solid or liquid matter, such as soot, dust, fumes, or mist.

■ **partition chromatography**

separation process in which one of the liquid phases is held stationary on a solid support while the other is allowed to flow freely down the column. Solutes partition themselves between the two phases based on their individual partition coefficients. Liquid-liquid chromatography is an example.

■ **partition coefficient (k)**

the amount of solute in the stationary phase relative to the amount of solute in the mobile phase. Can be distribution coefficicent, KD

■ **Pascal's law for pressure**

the basic property of a static fluid is pressure. Pressure is familiar as a surface force exerted by a fluid against the walls of its container. Pressure also exists at every point within a volume of fluid. For a static fluid, pressure turns to be independent of direction.

■ **Pascal's wager**

a line of logic used by Pascal, the famous mathematician. Its general idea is that, if there is no God and one does not believe in him, nothing is lost. Likewise, one loses nothing to believe in God if He does not exist. But if God does exist and one does not believe in him, a great loss has been suffered. Likewise, if there is a God and one believes in Him, one gains all.

■ **Paschen series**

Paschen series are the series of lines in the spectrum of the hydrogen atom which corresponds to transitions between the state with principal quantum number $n = 3$ and successive higher states.

■ **passiflora incarnata**

passion flower extract, mild fruit acids, help exfoliate and cleanse the skin.

■ **patch**

for corrosion modeling, one of two geometries for an opening in a waste package layer created by corrosion (the other geometry is a pit). A patch is generally wider than it is deep.

■ **patch test**

describes the technique of applying a small amount of a product prior to full use to see if any allergic reaction occurs. The best places to try a patch test are on the inside of the elbow or, if that is not appropriable, the inside of the wrist. Leave at least two hours for each test. Doing a patch test is strongly advisable prior to buying/trying a new product for anyone with sensitive skin.

■ **patter area**

the designed configuration area which includes the pattern and background. The pattern area bounds can be defined by a physical outline or by an imaginary outline formed by enclosing the pattern inside a box.

■ **Pauli exclusion principle**

Pauli exclusion principle is the statement that two electrons in an atom cannot have identical all four quantum numbers. It was first formulated in 1925 by Austrian-born Swiss physicst Wolfgang Ernst Pauli (1900-1958).

■ PCB

polychlorinated biphenyls; polychloro-biphenyls. Difficult to remediate chemical used in old-style transformers. Concentrated PCBs used to be referred to as '1268'.

■ p-doping

when Silicon is doped with a group III element

■ peak shape

describes the profile of a chromato graphic peak. Theory assumes a Gaussian peak shape (perfectly symmetrical); peak asymmetry factor describes shape as a ratio.

■ peak width

same as Band width

■ pearlite

a two-phase microctructure found in some steels and cast irons. It results from the transformation of austenite of eutectoid compositions and consists of alternating layers of alpha-ferrite and cementite.

■ pei

polyethyleneimine, an anionic polymeric phase used to coat silica. Most often used for the separation of proteins and peptides.

■ pellicular

see **porous-layer bead.**

■ peltier effect

peltier effect is the absorption or generation of heat (depending on the current direction) which occurs when an electric current is passed through a junction between two materials.

■ pendant group

side-group An offshoot, neither oligomeric nor polymeric , from a chain.

■ peptide

an organic compound containing the group -CONH (see below), examples of which are polypeptides.

■ poly

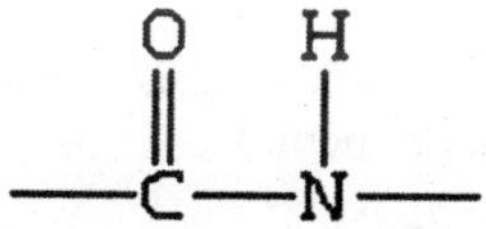

■ peptides

peptides are amides derived from two or more amino acids (the same or different) linked by peptide bonds. These bonds are formed by the reaction between adjancent carboxyl (-COOH) i amino (-NH_2) groups with the elimination of water.

■ perched water

groundwater of limited lateral extent separated from an underlying body of groundwater by an unsaturated zone.

■ percolating water

water passing through a porous substance. In rock or soil it is the movement of water through the interstices and pores under hydrostatic pressure and the influence of gravity. The downward or lateral flow of water that becomes net infiltration in the unsaturated zone.

■ **percolation**

the downward or lateral flow of water that becomes net infiltration in the unsaturated zone.

■ **percolation flux**

volumetric percolation rate per unit area. The flux anywhere below the root zone of plants and is no longer susceptible to removal back into the atmosphere by evapotranspiration. Volume of water moving downward or laterally through the unsaturated zone in a given period.

■ **perfect**

a perfect fluid is defined as a fluid with zero viscosity (i.e. inviscid).

■ **performance assessment**

an analysis that forecasts the behaviour of a system or system component under a given set of constant and/or transient conditions. Performance assessment includes estimates of the effects of uncertainties in data and modeling.

■ **periodic copolymer**

a copolymer consisting of macromolecules comprising more than two species of monomeric units in regular sequence.

■ **periodic copolymerisation**

a copolymerisation in which a periodic copolymer is formed.

■ **periodic table**

periodic table is a table of the elements, written in sequence in the order of atomic number or atomic weight and arranged in horizontal rows (periods) and vertical columns (groups) to illustrate the occurrence of similarities in the properties of the elements as a periodic function of the sequence. The original form was proposed by Dmitri Mendeleev (1834-1907) in 1869 using relative atomic masses.

■ **periods**

periods are horizontal rows in the periodic table. Each period beginning with an alkali metal (one electron in the outermost principal quantum level) and ending with an noble gas (each having eight electrons in the outermost principal quantum level, except for helium, which is limited to two).

■ **peristaltic pump**

a type of positive displacement pump.

■ **permanent gas**

a gas that cannot be condensed (or solidified) under the range of conditions of interest (e.g. temperature or pressure).

■ **permeability**

in general terms, the capacity of a medium (like rock, sediment, or soil) to transmit liquid or gas.

■ **permeation**

in SEC, refers to the process in which a solutc can enter a mobile-phase-filled pore of the packing.

permeation rate

the Permeation Rate is a measure of the rate at which a chemical will pass through protective material, such as that used for gloves. It is generally specified as the mass of material passing through unit area in unit time. For this value to be meaningful, the thickness of the protective material must also be specified.

permittivity

the proportionality constant between the dielectric displacement D and the electric field E.

peroxides

peroxides are compounds of structure ROOR in which R may be any organic group. In inorganic chemistry, salts of the anion O_2^{-2}. They are strong oxidising agents.

peroxidizable materials

peroxidizable materials can form peroxides in storage, generally when in contact with the air. These peroxides present their most serious risk when the peroxide-contaminated material is heated or distilled, but they may also be sensitive to mechanical shock. The quantity of peroxides in a sample may be determined using a simple peroxide test strip.

perpendicularity

a measure by the degree to which the angle between the x-x AND y-y axis approaches a right angle.

petrochemical

an intermediate chemical derived from petroleum, hydrocarbon liquids or natural gas, such as: ethylene, propylene, benzene, toluene and xylene.

petroleum ether

petroleum ether is the petroleum fraction consisting of C_5 and C_6 hydrocarbons and boiling in the range 35 °C to 60 °C; commonly used as a laboratory solvent.

petroleum jelly

(Petrolatum, Vaseline) A salve-like material derived from petroleum, which is used as a lubricant and emollient in many cosmetics and toiletries. It is not considered toxic. Sensitisation is possible and it can hinder normal skin respiration by keeping oxygen out.

petroleum

a generic name for hydrocarbons, including crude oil, natural gas liquids, natural gas and their products.

PFR

plug flow reactor. This may be thought of as a long pipe to which the reactants flow in, and the products flow out. In an ideal PFR, the components will be distributed axially, but will have uniform radial distribution.

pH

pH is a convenient measure of the acid-base character of a solution, usually defined by

$pH = -\log c(H^+)$

where $c(H^+)$ is the concentration of hydrogen ions in moles per liter. The more precise definition

is in terms af activity rather than concentration.
A solution of pH 0 to 7 is acid, pH of 7 is neutral, pH over 7 to 14 is alkaline.

■ phase

phase is a portion of a physical system (solid, liquid, gas) that is homogeneous throughout, has definable boundaries, and can be separated physically from other phases.

■ phase diagram

phase diagram is a graphical representation of the equilibrium relationships between phases (such as vapour-liquid, liquid-solid) of a chemical compound, mixture of compounds, or solution.

■ phase grating

what is recorded in the hologram by means of the photorefractive effect when two light beams interfere.

■ phase margin

the difference in phase at the frequency where the combined process and controller amplitude ratio is 0 is the phase margin.

■ phase stability

a measure of the ability of matter to remain in a given phase.

■ phase transformation

a change in the number and/or character of the phases that constitute the microstructure of an alloy.

■ phenol

an organic substance in which a hydroxyl group is attached directly to a benzene ring. An example is vanillin.

■ phenotype

the physical manifestation of the genes of an individual

■ phenoxyethanol

an antimicrobial agent (preservative) often used in combination with parabens. It is manufactured by combining phenol (coal tar derived) with ethylene oxide. It is irritant at high concentrations and sensitisation is possible.

■ phenyl phase

a nonpolar bonded phase prepared hy the reaction of dimethylphenylchlorosilane with silica gel. Claimed to have affinity for aromatic compounds.

■ phonon

a single quantum of vibrational or elastic energy.

■ phosphates

a substance that is added to a detergent to increase its water softening ability.

■ phosphoric acid

used for pH adjustments of products. Irritant in concentration, it is not considered toxic in cosmetic use. (used in coca-cola)

■ photo chemical machining, electro chemical milling process definitions

the precision removal from a thin metal sheet by etching or the conversion of material by electrochemical reaction. The chemmachining or milling process results in a very high tolerance with the parts being fabricated.

■ **photo etching fabrication (photofabrication)**

the production of precise metal shapes and other substances by reproducing a precise photographic image on the surface of the substance & etching away (or depositing upon) the unprotected areas by chemmachining or electrical means.

■ **photo mask definition**

an accurately made phototool whose images may be transferred by light exposure into a photoresist film coating.

■ **photo plotting**

photographic procedures whereby an image is generated by a controlled light's beam motion and/or positioning by numerical control directly onto a light sensitive material (usually emulsion).

■ **photo tool**

a general term applicable to any specifically defined photographic products. The artwork, artwork master, working master, and photomasks are each individually phototools.

■ **photochem, photoetch, photographic etching, photomill, photomilling, photomachining, process definitions**

the precision removal from a thin metal sheet by etching. The photochem, photoetch, photoetching, photomill, photo etching / milling, or photomachining process results in a very high tolerance with the parts being fabricated.

■ **photochemical decomposition**

the process of breaking apart stable molecules by ultraviolet radiation.

■ **photochemical reaction**

a chemical reaction can be enhanced when induced by light, just as some are speeded up by catalysts. *Chlorine* is one of a number of substances that undergo photochemical reactions. In sunlight, *chlorine* will react with *carbon monoxide* to produce the poisonous gas *phosgene*:

■ **photo 1**

■ **photocycle**

or photochemical cycle; the cycle of isomers and other slightly structurally different states of bacteriorhodopsin that occurs as a consequence of light striking it

■ **photoelectric**

a substance is photoelectric if it ejects electrons when light strikes it.

■ **photoelectric effect**

photoelectric effect is the complete absorption of a photon by a solid with the emission of an electron.

■ **photographic fog**

any plus density on negative working photographic products or a density loss on positive working products that appear on a exposed processed glass, film, or paper, but is not the result of image exposure. Fog can be produced by numerous causes including safelight, chem, diachronic, & fogs caused by aerial oxidation.

■ photographic layers

a photo-sensitive coating containing silver halide, which has been exposed & treated to yield a visible image of dispersed metallic silver obtained by photographic processes.

■ photoisomer

a molecule that differs from another in that one of its double bonds has been broken, the piece attached there twisted and the bond reformed. This requires considerably more energy than a conformation change, in that a bond must be broken for an isomer. A photoisomer would be an isomer that is caused by light.

■ photometry

visible light intensity, energy measurement as it affects the human eye.

■ photon

all matter absorbs and emits electromagnetic radiation in discrete small quantities called photons. The energy of a photon (E) varies withthe frequency of the radiation according to:

$E = hv$

- where h is a constant known as Planck's constant and v is frequency.

■ photoresist hardening

the process whereby photoresist is rendered more resistant to the plating effects and/or etching operations.

■ photoresist integrity

the photoresist soundness measured using such terms as chemical resistance, uniformity & adhesion.

■ photoresist

a material which when properly applied to a substrate becomes sensitive to portions of the electromagnetic spectrum. When properly exposed and developed, substrate portions are masked with a high degree of integrity.

■ photosensitive

the term used to describe a material that undergoes a change when light, or a specific wavelength of light, strikes it.

■ photosynthesis

phototsynthesis is the process by which organisms make their own food from sunlight.

■ physical properties

qualitative and Quantitative properties that describe a substance. They include smell, taste, colour, melting point, density, hardness etc.

■ phytoplankton

passively floating plant life in a body of water.

■ phytosterols

fatty alcohols that are derived from plants

phytotherapy

the use of plants and herbs for their therapeutic value.

picocurie

one trillionth of a curie (3.7×10^{-2} disintegrations per second or 2.22 disintegrations per minute), abbreviated pCi.

pid controller

controllers are designed to eliminate the need for continuous operator attention. Cruise control in a car and a house thermostat are common examples of how controllers are used to automatically adjust some variable to hold the process variable (or process variable) at the set-point. The set-point is where you would like the process variable to be. Error is defined as the difference between set-point and process variable.

(error) = (set-point) - (process variable)

The output of PID controllers will change in response to a *change* in process variable or set-point.

piezoelectric

a dielectric material in which polarisation is induced by the application of external forces.

piezometer

for measuring pressure inside a vessel or pipe in which liquid is there, a tube may be attached to the walls of the container (or pipe) in which the liquid resides so liquid can rise in the tube. By determining the height to which liquid rises and using the relation P_1 = ñgh, gauge pressure of the liquid can be determined. Such a device is known as *piezometer*. To avoid capillary effects, a piezometer's tube should be about 1/2 inch or greater. It is important that the opening of the device to be tangential to any fluid motion, otherwise an erroneous reading will result.

pigment

finely ground, natural or synthetic, inorganic or organic, insoluble dispersed particles (powder) which, when dispersed in a liquid vehicle to make paint, may provide, in addition to colour, many of the essential properties of the paint: opacity, hardness, durability, and corrosion resistance. The term is used to include extenders, as well as white or colour pigments. The distinction between powders which are pigments and those which are dyes is generally considered to be on the basis of solubility. Pigments being insoluble and dispersed in the material, dyes being soluble or in solution when used.

pine oil

an oil process from gum of pine trees.

pirkle columns

chiral 'brush-type' stationary phases, based on 3, 5-dinitrobenzoylphenylglycine silica, used in the separation of a wide variety of enantiomers. Named after the developer, Dr. William Pirkle, University of Illinois.

■ **p-isoamyi methoxycinnamate**

a UV-filter derived from cinnamic acid esters.

■ **pit**

for corrosion modeling, one of two geometries for an opening in a waste package layer created by corrosion (the other geometry is a patch). A pit is generally deeper than it is wide.

■ **pk**

pK is the negative logarithm (base 10) of an equilibrium constant K.

■ **planck constant**

planck constant (h) is a constant that when multipled by the frequency of radiation gives the quantity of energy contained in one quantum.

$E = hi$

Equal to $6.6260755(40) \cdot 10^{-34}$ J s. It is named after Max Planck (1858-1947).

■ **plasma**

plasma is a highly ionised gas in which the charge of the electrons is balanced by the charge of the positive ions, so that the system as a whole is electrically neutral. Temperatures of 10 000 °C to 15 000 °C can be reached.

■ **plasmid**

a loop of DNA in bacteria which floats in the cytoplasm.

■ **plastic**

a solid material in the primary ingredient of which is an organic polymer of high molecular wight.

■ **plastic deformation**

deformation that is permanent or nonrecoverable after release of the applied load.

■ **plasticiser**

a low molecular weight polymer additive that enhances flexibility and workability and reduces stiffness and brittleness.

■ **plate height (h)**

see **hetp**.

■ **plates**

refers to theoretical plates in a packed column.

■ **PLC**

Programmable Logic Controller. These computers replace relay logic and usually have PID controllers built into them. PLCs are very fast at processing discrete signals (like a switch condition). The most popular PLC manufacturer's are Allen Bradley, Modicon, GE, and Siemens (or TI).

■ **point defect**

a crystalline defect associated with one or, at most, several atomic sites.

■ **poise**

cgs unit of absolute viscosity. This is the shear stress (in dynes per square) required to move one layer of fluid along another over a total layer thickness of one centimeter at a shear rate of one centimeter per second. Dimensions are dyne-sec/cm². The Centipose (cP) is 1/100 of a poise and is the unit of absolute viscosity most commonly used. Whereas ordinary viscosity measurements depend on the force of gravity on the fluid to supply the shear stress and are thus subject to distortion by differences in fluid density. Absolute Viscosity measurements are independent of density and are directly related to resistance to flow.

■ poisson's ratio

for elastic deformation, the negative ratio of lateral and axial strains that result from an applied axial stress.

■ polar molecule

a molecule in which there exists a permanent electric dipole moment by virtue of the asymmetrical distribution of positively and negatively charged regions.

■ polar solvent

water is the most common polar solvent.

■ polarisation (electronic)

for an atom, the displacement of the centre of the negatively charged electron cloud relative to the positive nucleus, which is induced by an electric field. -

■ polarisation (ionic)

polarisation as a result of the displacement of anions and cations in opposite directions.

■ polarisation (orientation)

polarisation resulting from the alignment (by rotation) of permanent electric dipole moments with an applied electric field.

■ pollution

the impairment (reduction) of water quality by agriculture, domestic or industrial wastes (including thermal and radioactive wastes) to such a degree as to hinder any beneficial use of the water or render it offensive to the senses of sight, taste, or smell or when sufficient amounts of waste creates or poses a potential threat to human health or the environment.

■ polyacrylamide

a polymer produced by combination of acrylamide and sodium acrylate that is used as a thickener and additive in tanning products and nail polish. It is toxic and irritant.

■ polyacrylamide gel

neutral hydrophilic polymeric packings used in aqueous SEC. Prepared by the copolymerisation of acrylamide with (N,N'-methylene) bisacrylamide.

■ polyaddition

a polymerisation in which the growth of polymer chains proceeds by addition reactions between molecules of all degrees of polymerisation.

■ polycondensation

a polymerisation in which the growth of polymer chains proceeds by condensation reactions between molecules of all degrees of polymerisation.

■ polycrystalline

referring to crystalline materials that are composed of more than one crystal or grain.

■ polyelectrolyte

a polymer composed of polyelectrolyte molecules.

■ polyelectrolyte molecule

a macromolecule in which a substantial portion of the constitutional units have ionisable or ionic groups, or both.

■ polymer

polymer is a substance composed of molecules of high relative molecular mass (molecular weight), the structure of which essentially comprises the multiple repetition of units derived, actually or conceptually, from molecules of low relative molecular mass (monomers). In most cases the number of monomers is quite large and often is not precisely known. A single molecule of a polymer is called a macromolecule.

■ polymer blend

a macroscopically homogeneous mixture of two or more different species of polymer . In most cases, blends are homogeneous on scales smaller than several times visual optical wavelengths. For polymer blends, no account is taken of the miscibility or immiscibility of the constituent polymers, i.e., no assumption is made regarding the number of phases present. The use of the term *polymer alloy* for a polymer blend is discouraged.

■ polymeric packings

packings based on polymeric materials, usually in the form of spherical beads. Common polymers used in LC are polystyrene-divinylbenzene. polyacrylamide, polymethylacrylate, polyethylene-oxide, polydextran, and polysaccharide .

■ polymerisation

process of combining two or more simple molecules of the same type, called monomers, to form a single molecule having the same elements in the same proportion as in the original molecules, but having increased molecular weight. The product of the combination is a polymer.

■ polymer-polymer complex

a complex, at least two components of which are different polymers.

■ polymers

giant molecules in which between one hundred and one million subunits (monomers) are linked together. Polymers can be designed using specific monomers so that it is possible to plan and influence their characteristics. Special polymers - polycarboxylates - are used to reduce water hardness.

■ polymorphism

the ability of a solid material to exist in more than one form or crystal structure.

■ polypeptides

polypeptides are peptides containing ten or more amino acid residues. The properties of a polypeptide are determined by the type and sequence of its constituent amino acids.

■ polyquaternium companiol

a synthetic polymer that is combined with ammonium to produce a positively charged nitrogen, rendering it a cationic surfactant. They are used in hair-care products because of their sub-stantivity to hair.

■ polysaccharides

polysaccharides are compounds consisting of a large number of monosaccharides linked together by glycosidic bonds. Some important examples are starch, glycogen, and cellulose.

■ polysorbates 1-85

products of the combination of lauric acid with sorbitol, which is condensed with ethylene oxide to render it water-soluble. They are generally considered mild and non-toxic. Their main functions are as solubilisers and emulsifiers.

■ polystyrene-divinylbenzene resin

the most common polymer base for ion-exchange chromatography. Ionic groups are incorporated by various chemical reactions. Neutral PS-DVB beads are used in reversed-phase chromatography. Porosityand mechanical stability can be altered by varying the cross-linking through the variation of the DVB content.

■ pore diameter

same as mean pore diameter.

■ pore volume

the total volume of the pores in a porous packing; usually expressed in mL/g. It is measured by the BET method of nitrogen adsorption or by mercury intrusion, where Hg is pumped into the pores under high pressure.

■ pore water

the water and any material it is carrying that exist in the pore spaces of the rock matrix. Also pore fluid.

■ porosity

for a porous adsorbent, the ratio of the vol ume of the interstices to the volume of the solid particles. The pore volume is also used as a measure of porosity.

■ porous

a surface that was many tiny openings. A porous surface will require more finish or sealer to fill and smooth out these openings.

■ porous-layer bead

a small glass bead coated with a thin layer of stationary phase. The thin layer can be an adsorbent, a resin, or a phase chemically bonded onto the adsorbent. These were the first packings to be used in HPLC. They were larger (20-40 μm) than the microparticulate packings of today, but were easy to pack into columns and gave adequate efficiency. Also referred to as controlled surface porosity supports and pellicular packings.

■ **positron**

positron is the antiparticle of the electron. It has the same mass and spin as an electron, and an equal but opposite charge.

■ **positron emission tomography**

tomography produced by the detection of gamma rays emitted from tissues after administration of a natural biochemical substance into which positron-emitting isotopes have been incorporated; the paths of the gamma rays are interpreted by a computer, and the resulting tomogram represents local concentrations of the isotope-containing substance; used for the in vivo study of human metabolic and physiological functions.

■ **postclosure**

the period of time after closure of the geologic repository.

■ **post-synaptic cell**

neuron receiving a message in the form of an electrical impulse.

■ **potentiometric**

pertaining to the distribution of groundwater level.

■ **poultice**

plant material that is prepared in a special way and applied to the surface of the body as a remedy for certain disorders.

■ **pour point**

it is widely used low-temperature flow indicator and is 3°C above the temperature to which a normally liquid petroleum product maintains fluidity. It is a significant factor is cold-weather start-up, but must be considered along with pumpability, the ease with which forms a honeycomb of crystals at low temperatures near the pour point. However, agitation by a pump breaks down this wax structure and allows paraffinic oil to be pumped at temperatures well below their pour point. Naphthenic oils, on the other hand, contain little or no wax and reach their pour point through increase in viscosity: they cannot be pumped readily near the pour point.

■ **ppm**

parts per million - the unit commonly used to designate the concentration of a substance in a wastewater in terms of weight ie. one pound per million pounds, etc. PPM is synonymous with the more commonly used term mg/L (milligrams per liter).

■ **precipitate**

a solid particle that has separated from a liquid as a result of physical or chemical changes.

■ **precipitate**

material settled out of solution.

■ **precipitation**

1. the process of substance coming out of solution by the action of gravity or by a chemical reaction.
2. any form of water particles, such as frozen water in snow or

ice crystals, or liquid water in raindrops or drizzle, that falls from clouds in the atmosphere and reaches the earth's surface.

3 an amount of water that has fallen at a given point over a specified period of time, measured by a rain gauge.

■ precipitation hardening

hardening and strengthening of a metal alloy by extremely small and uniformly dispersed particles that precipitate from a supersaturated solid solution.

■ preclosure safety evaluation

a preliminary assessment of the adequacy of repository support facilities to prevent or mitigate the effects of postulated initiating event sequences and their consequences and the site, structures, systems, components, equipment, and operator actions that would be relied on for safety.

■ precolumn

a small column placed between the pump and the injector. It removes particulate matter that may be in the mobile phase, chemically sorbs substances that might interfere with the separation. or, as a saturator column, presaturates the mobile phase with stationary phase to prevent stripping of the column. Its volume has little effect on isocratic elution but contributes a delay to the gradient in gradient elution .

■ precursor

the reactants used in industrial semiconductor manufacture to make the desired product; for instance, ArH_3 is often the precursor for the Arsenic in GaAs devices

■ pre-exponential factor

also known as the frequency factor. Constant A (also denoted as Z, k^*, k_o) in the Arrhenius equation. The pre-exponential factor is associated with the frequency of collisions between molecules (entropy) and with the probability that these collisions result in a reaction.

■ preparative chromatograpby

the process of using liquid chromatography to isolate a sufficient amount of material for other experimental or functional purposes. For pharmaceutical or biotechnological purifications, columns several feet in diameter can be used for multiple grams of material. For isolating just a few micrograms of a valuable natural product, an analytical column can be used. Both are preparative chromatographic approaches.

■ pre-polymer

a polymer or oligomer composed of pre-polymer molecules.

■ pre-polymer molecule

a macromolecule or oligomer molecule capable of entering, through reactive groups, into further polymerisation, thereby contributing more than one monomeric unit to at least one chain of the final macromolecule. A pre-polymer molecule capable of entering into further polymerisation through reactive end-groups, often deliberately introduced, is known as a telechelic molecule.

■ **prepreg**

continuous fiber reinforcement pre-impregnated with a polymer resin which is then partially cured.

■ **preservatives**

floor finishes are susceptible to bacterial contamination. This is why finishes contain small amounts of antimicrobial agents to prevent microbial deterioration. These preservatives protect the unopened container, but do not substantially protect finish after it has been used. This is why it is important to never pour used floor finish back into a container of unused finish.

■ **presintering**

in powder metallurgy, a preliminary sintering operation performed at a temperature lower than the final sintering temperature.

■ **pressure**

pressure (p) is the force acting normally on unit area of a surface. It is measured in pascals in SI units (Pa)

$p = F/A$

■ **pressurised water reactor**

a type of nuclear power reactor that uses uranium fuel elements cooled and moderated by water under high pressure to keep the water from boiling. The water boiled to generate steam is in an external heat exchanger rather than in the reactor vessel body.

■ **pre-synaptic cell**

neuron attempting to transmit a message across a synapse in the form of an electrical impulse.

■ **preventive measures**

measures taken to avoid further development of a runaway or to reduce its final effects.

■ **primary bond**

interatomic bonds that are relatively strong and for which bonding energies are relatively large.

■ **probabilistic analyses**

analyses in which uncertainty in processes and events is represented through probability distributions for the parameters of those processes and events.

■ **probabilistic risk assessment**

1. a systematic process of identifying and quantifying the consequences of scenarios that could cause a release of radioactive materials to the environment.
2. using predictable behaviour to define the performance of natural, geologic, human, and engineered systems for thousands of years into the future using probability distributions.

■ **probability**

the expression for the likelihood of occurrence of an event or event sequence during an interval of time or the likelihood of a success or failure of an event on test or demand. By definition, probability is expressed as a number ranging from 0 to 1.

■ **process model**

a depiction or representation of a process along with any hypotheses required to describe or to explain the process.

■ **processor**

the processor in a computer is the main chip. Often it is called the central processing unit, or CPU. PowerPC, Pentium, and 80486 are common processors. Processor speed is measured in megahertz, or millions of cycles per second. Common speeds are 133, 150, 180 and 225 megahertz, or MHz.

■ **products**

chemicals produced during a reaction process.

■ **propane**

an inflammable propellant gas used in aerosols.

■ **propellants**

these are pressurised gases used to propel the contents of aerosol sprays from the can, through the nozzle giving a fine mist. Until recently CFC's were widely used in this role, but they have now been replaced with less damaging chemicals. Most propellants contribute to global warming and are often highly flammable, e.g. Butane.

■ **propolis**

a substance produced by bees to protect their hive from viral or bacterial attack. Propolis has a preservative and antibiotic action. It is available as a tincture and may be combined into cosmetic formulations.

■ **proportional band**

the 'P' of PID controllers. With proportional band, the controller output is proportional to the error or a change in process variable. Proportional Band = 100/ Gain.

■ **proportional gain**

this is the 'P' part of the PID controller.

■ **proportional limit**

the point on a stress-strain curve at which the straight line proportionality between stress and strain ceases.

■ **propyl parabens**

a commonly used broad-spectrum preservative. Recent evidence suggests that people with certain types of cancer produce an excess of propyl parabens and due to this link the popularity of this substance is declining. There is no evidence to date that propyl parabens causes cancer. See also Parabens.

■ **propylene**

glycol A colourless, almost odourless liquid, synthetically produced. It is used in perfumes and flavourings and as a solvent, humectant and moisturising ingredient in a wide range of cosmetics and toiletries. It is a strong skin irritant and can cause liver abnormalities and kidney damage, along with eye irritation, nausea and headaches.

■ **prosthesis, prostheses**

an artificial device used to replace a missing body part, such as a limb, a tooth, an eye, or a heart valve. Some are mere attachments, while sensory prostheses respond to the electrical impulses sent to muscles and allow for limited control of an artificial device for actions such as raising and arm and contracting and opening an artificial hand.

■ **protein**

building blocks of the human body, which are composed of a combination of amino acids. Keratin, which makes up the hair, is an example.

■ **proton**

proton is a stable elementary particle of unit positive charge and spin 1/2. Protons and neutrons, which are collectively called nucleons, are the constituents of the nucleus.

■ **pseudocolloid**

a colloid from natural or man-made materials, as distinguished from a colloid from insoluble radioactive materials (intrinsic colloid) or from altered fragments of spent nuclear fuel or glass waste forms (waste-form colloid).

■ **pseudo-co-oligomer**

an irregular oligomer , the molecules of which are derived from only one species of monomer but which display a variety of structural features more appropriate for description in co-oligomer terms.

■ **pseudo-copolymer**

an irregular polymer , the molecules of which are derived from only one species of monomer but which display a variety of structural features more appropriate for description in copolymer terms. Where appropriate, adjectives specifying the types of *copolymer* may be applied to *pseudocopolymer*. The term statistical pseudo-copolymer, for instance, may be used to describe an irregular polymer in the molecules of which the sequential distribution of configurational units obeys known statistical laws.

■ **pseudo-plastic fluid**

most non-Newtonian fluids fall into this group. Viscosity decreases with increasing velocity gradient. e.g. polymer solutions, blood. Pseudoplastic fluids are also called as Shear thinning fluids. At low shear rates(du/dy) the shear thinning fluid is more viscous than the Newtonian fluid, and at high shear rates it is less viscous.

■ ***p* type semiconductor**

a semiconductor for which the predominant charge carriers responsible for electrical conduction are holes. Normally, acceptor impurity atoms give rise to the excess holes.

■ **pulsating flow**

flow originating from a reciprocating pump. Normally, a pulse damper dampens the pulses out, by an electronic pressure feedback circuit, or by an active damper pump head. Some detectors (e.g., electrochemical) are affected by flow pulsations.

pungent

irritating or sharply painful. Producing a sharp sensation of taste or smell.

purgative

a substance that promotes the vigorous evacuation of the bowels. Usually used to relieve severe constipation. Examples: Aloe, Epsom salt, licorice, rhubarb, safflower, senna.

purifier

a machine used for a liquid-liquid separation in which the two intermixed liquids which are insoluble in each other have different specific gravities. Solids with specific gravities higher than those of the liquids can be separated off at the same time. A purifier bowl has two outlets; one for the light phase liquid and one for the heavy phase liquid.

pv or process variable

what you are trying to control temperature, pressure, flow, composition, pH, etc. Also called the measurement.

PVA (Polyvinyl Acetate)

a colourless, thermoplastic, water soluble, resinous high polymer derived from the polymerisation of vinyl acetate with a catalyst; used as a latex binder in certain, generally lower quality water-base coatings.

PVC (Pigment Volume Concentration)

the ratio of the volume of pigment to the volume of total nonvolatile material (i.e., pigment and binder) present in a coating. The figure is usually expressed as a percentage.

pyroclastic

of or relating to clastic rock material of any size that is formed by volcanic explosion or ejected from a volcanic vent.

pyrolysis

chemical change caused by heat.

pyrophoric materials

pyrophoric materials ignite spontaneously in air. Since a wide variety of chemicals will burn if heated sufficiently, it is usual to define a pyrophoric material as one which will ignite spontaneously at temperatures below about 45C.

qualitative analysis

qualitative analysis involves determining the nature of a pure unknown compound or the compounds present in a mixture.

quantal effect

a quantal effect is one for which there are only two possible outcomes; the effect thus occurs or does not. In the field of safety, the context in which the term quantal effect is most widely used (and the most dramatic...) is death.

■ quantitative analysis

quantitative analysis is the determination of the amount of substances present in a sample.

■ quantum

quantum is the smallest quantity of energy that can be emitted (or absorbed) in the form of electromagnetic radiation. Energy of a quantum (E) is equal to

$E = h\nu$

where h is Planck constant and ν(ν→*nu*) is the frequency of the radiation.

■ quantum mechanics

a mathematical description of the behaviour of electrons. A description of the behaviour of electrons and the atomic world which is used to understand their quantised, or discrete properties (such as atomic line spectra).

■ quasicrystal

quasicrystal is a solid having conventional crystalline properties but whose lattice does not display translational periodicity.

■ quaternary ammonium compound

a compound based on ammonium salts where the hydrogen atoms have been replaced by other chemical groups. These compounds are cationic and can be combined with other materials to provide a range of chemical functions, mainly preservatives and surfactants. Examples are Quaternarium 1-30 and Benzalkonium chloride. Many are irritant in concentration. Sensitisation is possible.

■ quenching

rapid cooling of the reaction system in a short time (almost instantaneously). This condition 'freezes' the status of a process and prevents further reaction or decomposition.

■ racemate

racemate is a mixture of equal quantities of the *d*- and *l*-forms of an optically active compound. A racemic mixture is not optically active.

■ radial compression

the use of radial pressure applied to a flexible-wall column to lessen wall effects.

■ radial diffusion

diffusion across the LC column in a radial direction. If the sample is injected into the exact centre of a column, it will spread not only in a longitudinal direction as it flows, but radially as well.

■ radical

a radical is a reaction intermediate where a bond is broken and the two parts of the molecule exist carrying a single unshared electron. The radicals are normally destroyed quickly by recombination or they break in another radical and a stable molecule.

■ radical copolymerisation

a copolymerisation which is a radical polymerisation.

■ **radical polymerisation**

a chain polymerisation in which the kinetic-chain carriers are radicals.

■ **radioactive decay**

the process in which one radionuclide spontaneously transforms into one or more different radionuclides called decay products or daughter products.

■ **radioactive waste**

high-level radioactive waste and other radioactive materials, including spent nuclear fuel, that are received for emplacement in the geologic repository.

■ **radioactivity**

the property possessed by some elements (e.g., uranium) of spontaneously emitting alpha, beta, or gamma rays by the disintegration of atomic nuclei.

■ **radiologically controlled area**

an area of the surface repository enclosed by security fences, control gates, lighting, and access detection systems. This area includes the facilities and transportation systems required to receive and ship rail and truck waste shipments, prepare shipping casks for handling, and load waste forms into disposal containers for underground emplacement. It also includes the facility and systems required to treat and package site-generated, low-level radioactive waste for offsite disposal.

■ **radiolysis**

the chemical dissociation of molecules caused by exposure to radiation. For example, under certain circumstances, radiation can cause the hydrogen and oxygen molecules in water to separate.

■ **radionuclide**

a radioactive atom with an unstable nucleus that spontaneously decays, emitting ionising radiation in the process.

■ **radiosonde**

weather balloon.

■ **raise**

from mining terminology, an upward opening, either vertical or inclined, driven in rock from one level to that above it.

■ **ramsbottom coke:**

a carbon residue test originated by Dr. J.R. Ramsbottom in England

■ **random copolymer**

a copolymer consisting of macromolecules in which the probability of finding a given monomeric unit at any given site in the chain is independent of the nature of the adjacent units.In a random copolymer, the sequence distribution of monomeric units follows Bernoullian statistics.

■ **random copolymerisation**

a copolymerisation in which a random copolymer is formed.

■ **rankine cycle**

rankine cycle is a thermodynamic cycle which can be used

to calculate the ideal performance of a heat engine that uses a condensable vapour as the working fluid.

■ **raoult's law**

raoult's law is the expression for the vapour pressure p_A of component A in an ideal solution, viz.,

$p_A = p_A{}^o x_A$

where x_A is the mole fraction of component A and $p_A{}^o$ the vapour pressure of the pure substance A.

■ **rare earth elements**

rare earth elements are the elements Sc, Y, and the lanthanides (La, Ce, Pr, Nd, Pm, Sm, Eu, Gd, Tb, Dy, Ho, Er, Tm, Yb, Lu). These elements got their name from the fact that chemists first isolated them in their oxide forms. These oxides somewhat resemble calcium, magnesium and aluminum oxides, sometimes called common earths.

■ **rate**

same as the derivative or 'D' part of PID controllers.

■ **rate constant**

the proportionality constant between the rate of a reaction and the concentrations of the species that appear in the rate law, raised to the appropriate powers.

■ **rate equation**

1. see **arrhenius rate equation**.
2. an expression relating the rate of a reaction to the centrations of species present in the reaction mixture.

■ **rate of reaction**

the rate at which the conversion of reactants takes place. The rate of reaction (r) is a function of concentrations (F(c)) and the reaction rate coefficient (k):
r = k.F(cA, cB,....,cX). The heat (q) produced by a reaction is a linear function of the rate of reaction which makes the rate of reaction a basic parameter in determining the required cooling capacity during all stages of the reaction process.

■ **Rayleigh scattering**

Rayleigh scattering is the scattering of light by particles which are much smaller than the wavelength of the light.

■ **reactants**

chemicals that are converted during the reaction process.

■ **reaction**

the process in which chemicals (reactants) are converted into other chemicals (products).

■ **reaction calorimeter**

a laboratory test apparatus for measuring thermal effects of chemical reactions or processes. For the purpose of this guidance, types of reaction calorimeters include isothermal, isoperibolic and adiabatic systems.

■ **reaction kinetics**

a set of equations and parameters that determine a reaction rate.

■ **reaction rate constant**

the constant k in the rate of reaction. The reaction rate constant is a strong function of temperature as represented by the Arrhenius equation.

■ **reaction sintering**

materials science. A process in which a compacted powder is heated to a temperature where sintering takes place simultaneously with a chemical reaction; for example silicon powder heated in the presence of nitrogen to produce silicon nitride.

■ **reagent**

a substance used to start a chemical reaction. In the laboratory, hydrochloric acid, sulfuric acid and sodium hydroxide are reagents.

■ **recharge**

the movement of water from an unsaturated zone to the saturated zone.

■ **recovery**

the amount of solute (sample) that elutes from a column relative to the amount injected. Most ohen used with protein separations in which proteins hang up on active sites of the packing in certain columns.

■ **recrystallisation**

the formation of a new set of strain-free grains within a previously cold-worked material; normally an annealing heat treatment is necessary.

■ **recycle**

the use of water or wastewater within (internally) a facility before it is discharged to a treatment system.

■ **recycling**

a technique in which the column effluent is recirculated onto the head of the column in an attempt to gain the advantage of extended column length. Can be carried out on a single column by passing the emuent back through the pump. An alternative technique uses two columns connected by a switching valve. Very seldom used in HPLC, and then only in size-exclusion chromatography.

■ **reduced crude oil**

crude oil that has undergone at least one distillation process to separate some of the lighter hydrocarbons. Reducing crude lowers its API gravity, but increases the handling safety by raising the flash point.

■ **reduced plate height *(h)***

used to measure efficiencies of columns. An HPLC column with an h value ≤ 2 is considered to be well-packed. $h = H/d_p$.

■ **reduced velocity (ν)**

along with the reduced plate height, is used to compare different chromatographic columns. It relates the solute diffusion coefficient (D_m) in the mobile phase to the particle size of the column packing (d_p). $V = d_p/D_m$.

■ **reducing agent**

a reducing agent may be defined in various ways, depending upon the context in which the phrase is used. In broad terms it is often

taken to mean a chemical which can act as an electron donor. Thus, in the reaction
$Zn^{2+}(aq) + 2Fe^{2+}(aq) = Zn(s) + 2Fe^{3+}(aq)$
the zinc is being reduced (gaining electrons) by reaction with the iron cations; the Fe^{2+} in this instance is acting as a reducing agent. This defintion is far more general than the one which follows if we regard 'reduction' as being just the removal of oxygen from a substance. A variety of substances can act as reducing agents. Hydrogen on its own is a reasonable reducing agent, but most readily-oxidisable species are also reducing agents.

■ reduction

the addition of hydrogen and the removal of oxygen - the oppposite to oxidation, but is now generalised to the inclusion of electrons to a substance by any means.

■ reference biosphere

the description of the environment inhabited by the reasonably maximally exposed individual. The reference biosphere comprises the set of specific biotic and abiotic characteristics of the environment, including but not necessarily limited to climate, topography, soils, flora, fauna, and human activities.

■ refinery

a plan used to separate the various components present in crude oil and convert them into usable products or feedstock for other processes.

■ reflection

deflection of a light beam at the interface between two media.

■ reflux

an operation in which vapour is produced, condensed and subsequently returned to the originating vessel.

■ refraction

bending of a light beam upon passing from one medium into another.

■ refractive index

the ratio of the velocity of light in a vacuum to the velocity of light in some medium.

■ refractory

a metal or ceramic that may be exposed to extremely high temperatures without deteriorating rapidly or without melting.

■ refrigerant

relieves fever and thirst. A cooling remedy. Lowers body temperature. Examples: Aloe, coriander, ginger, hibiscus, orange, lemon, licorice, sandalwood.

■ regeneration

returning the packing in the column to its initial state after gradient elution. Mobile phase is passed through the column stepwise or in a gradient. The stationary phase is solvated to its original condition. In ion-exchange chromatography, regeneration involves replacing ions taken up in the exchange process with the original ions that occupied the exchange sites. Regeneration can also refer to bringingback any column to its original state (e.g., the removal of impurities with a strong solvent).

regular macromolecule

a macromolecule, the structure of which essentially comprises the repetition of a single constitutional unit with all units connected identically with respect to directional sense.

regular oligomer molecule

an oligomer molecule, the structure of which essentially comprises the repetition of a single constitutional unit with all units connected identically with respect to directional sense.

regular polymer

a polymer composed of regular macromolecules, regular star macromolecules, or regular comb macromolecules . A polymer consisting of star macromolecules with arms identical with respect to constitution and degree of polymerisation is considered to be regular. Analogously, a polymer consisting of comb macromolecules with the sub-chains between the branch points of the main chain and the terminal subchains of the main chain identical with respect to constitution and degree of polymerisation and the sidechains identical with respect to constitution and degree of polymerisation is considered to be regular.

regulator

when a controller changes a process variable to move the process variable back to the setpoint, it is called a regulator.

relative atomic mass

relative atomic mass (A_r) is the ratio of the average mass per atom of the naturally occurring form of an element to 1/12 of the mass of nuclide ^{12}C. The term atomic weight is synonymous with relative atomic mass.

relative density

relative density is the ratio of the density of a substance to the density of some reference substance. For liquids or solids it is the ratio of the density (usualy at 20 °C) to the density of water at 4 °C. This quantity was formerly called specific gravity.

relative humidity

relative humidity is the ratio of the partial pressure of water vapour in air to the saturation vapour pressure of water at the same temperature, expressed as a percentage.

relative magnetic permeability

the ratio of the magnetic permeability of some medium to that of a vacuum.

relative molecular mass

relative molecular mass (M_r) is the ratio of the average mass per molecule or specified entity of a substance to 1/12 of the mass of nuclide ^{12}C. Also called molecular weight. It is equal to the sum of the relative atomic masses of all the atoms that comprise a molecule. For example

$M_r(H_2SO_4) = 2\times A_r(H) + A_r(S) + 4\times A_r(O)$
$= 2\times 1.0079 + 32.066 + 4\times 15.999$
$= 2.0158 + 32.066 + 63.996$
$= 98.078$

■ relative retention ratio

same as separation factor or selectivity.

■ rem

roentgen equivalent man. The unit of a dose equivalent from ionising radiation to the human body. It is used to measure the amount of radiation to which a person has been exposed.

■ reset

same as the integral or 'I' part of PID controllers.

■ reset windup

with a simple PID controller, integral action will continue to change the controller output value (in voltage, air signal or digital computer value) after the actual output reaches a physical limit. This is called reset (integral) windup. For example, if the controller is connected to a valve which is 100% open, the valve cannot open farther. However, the controller's calculation of its output can go past 100%, asking for more and more output even though the hardware cannot go past 100%. Most controllers use an 'anti-reset windup' feature that disables integral action using one of a variety of methods when the controller hits a limit.

■ residual silanols

the silanol (-Si-OH) groups that remain on the surface of a packing aher a phase is chemically bonded onto its surface. These silanol groups may not be accessible to the reacting bulky organosilane (e.g., octadecyldimethylchlorosilane) but may be accessible to small polar compounds. Often they are removed by endcapping with a small organosilane such as trimethylchlorosilane.

■ residual fuel oil

heavy fuel oils produced from the non-volatile residue from the fractional distillation process. Heavy oils that are 'leftovers' from various refining processes. Heavy black oils used in marine boilers and in heating plants.

■ residue

a single amino acid in the protein molecule.

■ resin

a solid packing used in ion-exchange separations. The most popular resins are polystyrene-divinylbenzene copolymers of small particle size (<10 m). Ionic functionality is incorporated into the resin.

■ resistance

resistance (*R*) is electric potential difference divided by current when there is no electromotive force in the conductor. This definition applies to direct current. More generally, resistance is defined as the real part of impedance.

■ **resistivity**

the reciprocal of electrical conductivity, and a measure of a material's resistance to the passage of electric current.

■ **resolution (rs)**

ability of a column to separate chromatographic peaks. It is usually expressed in terms of the separation of two peaks. One attempts to achieve the best resolution possible. Resolution can be calculated in two ways:
$R_s = 2(t_{R2} - t_{R1})/(W_{b,1} + W_{b,2})$
or $R_s = 1/4(a-1/a)\ (k'/1+k')N^{1/2}$
A value of 1 is considered to be the minimum for a measurable separation to occur and to allow good quantitation. Values of 1.7 or larger are generally desirable for rugged methods.

■ **resolvent**

promotes the resolving and removing of abnormal growths, such as a tumor.

■ **resonance**

a term used in spectroscopy that descibes a place where the energy of a transition for an atom or molecule is matched by the energy of a an electromagnetic wave. We can see where the resonances are by looking at a spectrum.

■ **resorcinol**

a synthetically derived alcohol that is used in dandruff shampoos to prevent itching; it also acts as a preservative and reduces skin greasiness. It is a known sensitiser and is irritant in concentration.

■ **retardation**

slowing of radionuclide movement in groundwater by mechanisms that include sorption of radionuclides, diffusion into rock matrix pores and microfractures, and trapping of large colloidal molecules or particles in small pore spaces or dead ends of microfractures.

■ **retention time (t_r)**

the time between injection andthe appearance of the peak maximum. The adjusted retention time t'_R adjusts for the column void volume. $t'_R = t_R - t_0$ (or t_m).

■ **retention volume (v_r)**

the volume of mobile phase required to elute a substance from the column $V_R = F \cdot t_R$ or $V_R = V_m - K_D V_s$, where Vm is the void volume, KD the distribution coefficient. and Vs the stationary phase volume

■ **reversed-phase chromatography**

the most common HPLC mode. Uses hydrophobic packings such as octadecyl- or octylsilane phases bonded to silica or neutral polymeric beads. Mobile phase is usually water and a water-miscible organic solvent such as methanol or acetonitrile. There are many variations of RPC in which various mobile phase additives are used to impan a different selectivity. For example, for the RPC of anions, the addition of a buffer and tetraalkylammonium salt would allow ion pairing to occur and effect separations that rival ionexchange chromatography.

■ reversible colloid

a pseudocolloid to which radionuclides are reversibly bound.

■ Reynolds number

Reynolds number (*Re*) is a dimensionless quantity used in fluid mechanics, defined by

$Re = \tilde{n}\, vl / \varsigma$

where $\tilde{n}$ is density, v is velocity, l is length, and ς is viscosity.

This is a number which characterises flow. If the Reynold's number is low (under approximately 1800) then the flow is said to be laminar. This may be thought of as the fluid flowing in layers. If the Reynold's number is high (over approximately 2300) then the flow is said to be turbulent. Turbulent flow is considered to be well mixed.

■ rheology

rheology is the study of the deformation and flow of materials. Has important bearing on the behaviour of viscous liquids in plastic molding.

■ rheopectic fluids

dynamic viscosity increases with the time for which shearing forces are applied. e.g. gypsum suspension in water.

■ rhodopsin

rhodopsin is a retinal photosensitive protein found in higher organisms, like mammals. It is a lot like bacteriorhodopsin, except that's found in bacteria and slightly different in structure.

■ ribwort juice

maintains healthy mucous membranes, useful for sinus problems.

■ ricinus communis

castor oil extract, derived from the seed of the Castor plant.

■ ring-opening copolymerisation

a copolymerisation which is a ring-opening polymerisation with respect to at least one monomer.

■ ring-opening polymerisation

a polymerisation in which a cyclic monomer yields a monomeric unit which is acyclic or contains fewer cycles than the monomer.

■ risk

the chance of something adverse happening where 'something' refers to a particular consequence of the manifestation of a hazard. Risk reflects both the likelihood that harm will occur and its severity in relation to the number of people who might be affected, and the consequences to them.

■ robust

a loop that is robust is relatively insensitive to process changes. A less robust loop is more sensitive to process changes.

■ rod-cone cells

photoreceptors in the eyes of vertebrate organisms; rod cells are sensitive to black and white and enable night vision; cone cells are sensitive to colour

■ **rodenticide**

a rodenticide is a term applied to any chemical used to kill rodents.

■ **rotational**

a rotational fluid flow can contain streamlines that loop back on themselves. Hence, fluid particles following such streamlines will travel along closed paths. Bounded (and hence nonuniform) viscous fluids exhibit rotational flow, typically within their boundary layers. Since all real fluids are viscous to some amount, all real fluids exhibit a level of rotational flow somewhere in their domain. Regions of rotational flow correspond to the regions of viscous losses in a fluid. Inviscid fluid flows can also be rotational, but these are special nonphysical cases. For an inviscid fluid flow to be rotational, it must be set up that way by initial conditions. The amount of rotation (called the *velocity circulation*) in an inviscid fluid flow is conserved, provided that the fluid is also barotropic and subject only to conservative body forces. This conservation is known as *Kelvin's Theorem* of constant circulation.

■ **runaway**

a reaction that is out of control because the rate of heat generation by an exothermic chemical reaction exceeds the rate of cooling available.

■ **runoff**

water from rain and snow that flows over land to streams.

■ **run-on**

the volume or depth of the routed surface-water flow.

■ **rupture**

failure that is accompanied by significant plastic deformation.

■ **rust inhibitor**

a lubricant additive for protecting ferrous (iron and steel) components from rusting caused by water contamination or other harmful materials formed by oil degradation. Some rust inhibitors operate similarity to corrosion inhibitors by reacting chemically to form an inert film on metal surfaces. Other rust inhibitors absorb water by incorporating it into water-in-oil emulsion so that only the oil touches the metal surfaces.

sadt (Self-Accelerating Decomposition Temperature)

the lowest ambient temperature at which auto-accelerative decomposition of an unstable substance is observed (minimum self-heating rate of 5 °C/week), when the substance as packaged in its commercial container and is subjected to that temperature during one week in the testing facility.

■ **safety case**

the set of data and analyses that, collectively, are intended to provide the reasonable assurance that a successful license application would require.

■ **salicylic acid**

this occurs naturally as methyl salicylate for example, in oil of wintergreen, from which it can be obtained by treatment with alcoholic potassium hydroxide. The synthetic version, which is more often used commercially, is used in cosmetics as an antiseptic and preservative and for treating acne. Sensitisation is possible.

■ **salinity**

salinity (S) is a parameter used in oceanography to describe the concentration of dissolved salts in seawater. The salinity of normal seawater is 35 parts salt per 1000 parts water.

■ **salt**

salt is an ionic compound formed by the reaction of an acid and a base. The reaction of sodium hidroxide to hydrochloric acid give sodium chloride

$NaOH + HCl$ à $NaCl + H_2O$

■ **salt bridge**

salt bridge is a permeable material soaked in a salt solution that allows ions to be transferred from one container to another. The salt solution remains unchanged during this transfer.

■ **salvia officinalis**

sage, antiseptic and astringent.

■ **sample capacity**

refers to the amount of sample that can be injected onto an LC column without overload. Ohen expressed as grams of sample per gram of packing. Overload is defined as the sample mass injected at which the column efficiency falls to 90% of its normal value.

■ **sample interval**

the rate at which a controller samples the process variable, and calculates a new output. Ideally, the sample interval should be set between 4 and 10 times faster than the process dead time.

■ **saturated**

describes a solution that will not dissolve any more solute at a given temperature. Any more solute will remain as crystals.

■ **saturated compound**

an organic compound that does not comprise of carbon-carbon multiple bonds. Examples include ethane and propane.

■ **saturated solution**

saturated solution is a solution that holds the maximum possible amount of dissolved material. When saturated, the rate of dissolving solid and that of recrystallisation solid are the same, and a condition of equilibrium is reached. The amount of material in solution varies with the temperature; cold solutions can hold less dissolved solid material than hot solutions. Gases are more soluble in cold liquids than in hot liquids.

■ **saturated zone**

the region below the water table where rock pores and fractures are completely saturated with water.

■ **saturator column**

see **precolumn.**

■ **SAX**

strong anion exchanger. A typical strong anion exchange functional group would be tetraalkylam monium .

■ **saybolt furol viscosity**

a viscosity test similar in nature to the Saybolt Universal viscosity test but one more appropriate for testing high=viscosity oils. Certain transmission and gear oils, and heavy fuel oils are rated by this method. The results obtained are approximately 1/ 10th the viscosity which would be shown by the Saybolt Universal method.

■ **scenario**

a well-defined, connected sequence of features, events, and processes that can be thought of as an outline of a possible future condition of the repository system. Scenarios can be undisturbed, reflecting performance under the expected, or nominal, behaviour for the system. The scenario can also be disturbed if altered by disruptive events such as human intrusion or natural phenomena such as volcanism.

■ **schiff base**

schiff base is a class of compounds derived by chemical reaction (condesation) of aldehydes or ketones with aromatic amines, for example

RNH_2 + R'CHO à RN:CHR' + H_2O

They are named after the German chemist Hugo Schiff (1834-1915).

■ **schrödinger equation**

schrödinger equation is the basic equation of wave mechanics which, for systems not dependent on time, takes the form: where is the wavefunction, V is the potential energy expressed as a function of the spatial coordinates, E its total energy, ∂^2 is the Laplacian operator, h is Planck's constant, and m is the mass.

■ **scientific method**

a standardised way that scientists research and find answers to questions and problems.

■ **scuffing**

engine wear resulting from the localised welding and fracture of rubbing surfaces.

■ **SCX**

strong cation exchanger. A typical strong cation exchange functional group would be a sulphonic acid.

■ **sd alcohol**

an alcohol that has been denatured by the addition of a compound that renders it undrinkable.

■ **second**

second (s) is the SI base unit of time.

The second is the duration of 9 192 631 770 periods of the radiation corresponding to the transition between the two hyperfine levels (F = 4, m_F = 0 to F = 3, m_F = 0) of the ground state of the caesium 133 atom.

■ **sedative**

herb that allays excitement, induces relaxation, and is conducive to sleep.

■ **sedimentation**

a technique used to size resins for ion-exchange chromatography. Particles of different size and density will settle at different velocities into a gradient. Very narrow cuts of particle size can be obtained.

■ **seepage**

1. the inflow of groundwater moving in fractures or pore spaces of permeable rock to an open space in the rock; the amount of percolation flux that enters the drift in a given time period.
2. flow of liquid water into an underground opening such as a waste emplacement drift or exploratory tunnel. Does not include water vapour movement into openings or condensation of water vapour within openings.

■ **seepage threshold**

a critical percolation flux below which seepage into the openings is unlikely to occur.

■ **selectivity (a)**

same as separation factor or relative retention ratio A thermodynamic factor that is a measure of relative retention of two substances. Fixed by a certain stationary phase and mobile phase composition. a = *k'2/k'1*.

■ **selenides**

selenides are compounds having the structure RSeR (R not equal to H). They are thus selenium analogues of ethers. Also used for metal salts of H_2Se.

■ **semiarid**

of a climate: having precipitation only sufficient for growth of sparse vegetation; a region in which the annual precipitation is about 250 to 500 mm (10 to 20 in.).

■ **semi-batch**

an operation in which some materials are added to the reactor at the start, with one or more other reactants added in a controlled manner during the reaction.

■ **semiconducting**

conducts electricity when a certain energy level is reached.

■ **semiconductor**

semiconductor is a material in which the highest occupied energy band (valence band) is completely filled with electrons at $T = 0$ K, and the energy gap to the next highest band (conduction band) ranges from 0 to 4 or 5 eV. With increasing temperature electrons are excited into the conduction band, leading to an increase in the electrical conductivity.

■ **Semi-Interpenetrating Polymer Network (SIPN)**

a polymer comprising one or more networks and one or more linear or branched polymer(s) characterised by the penetration on a molecular scale of at least one of the networks by at least some of the linear or branched macromolecules. Semi-interpenetrating polymer networks are distinguished from interpenetrating polymer networks because the constituent linear or branched polymers can, in principle, be separated from the constituent polymer network(s) without breaking chemical bonds; they are polymer blends.

■ semimetal

see **metalloid.**

■ semipreparative chromatography

refers to preparative liquid chromatography carried out on an analytical size (4-5 mm i.d.) or slightly larger (6-10 mm i.d.) column. Normal injection size is in the milligram to low-gram range.

■ sensitivity study

an analytic or numerical technique for examining the effects of varying specified parameters in a computer model. Shows the effects that changes in various parameters have on model outcomes and illustrates which parameters have a greater impact on the predicted behaviour of the system being modeled. Also called *sensitivity analysis* because it shows the sensitivity of the consequences (e.g., radionuclide release) to uncertain parameters (e.g., the infiltration rate that results from precipitation).

■ servo

when a controller changes a process variable to move the process variable in response to a setpoint change, it is called a servo.

■ set-point

the set-point is where you would like the process variable to be. For example, the room you are in now has a setpoint of about 70 degrees. The desired temperature you set on the thermostat is the setpoint.

■ short chain

a chain of low relative molecular mass.

■ short ton

an avoirdupois measure of weight equal to 2,000 lbs.

■ SI units

see **international system of units.**

■ siemens

siemens (S) is the SI derived unit of electrical conductance of a conductor in which a current of one ampere is produced by an electrical potential difference of one volt ($S = A \cdot V^{-1} = \grave{U}^{-1}$). The unit is named after the German scientist Ernst Werner von Siemens (1816-1892).

■ sievert

sievert (Sv) is the SI derived unit of dose equivalent when the absorbed dose of ionising radiation multiplied by the stipulated dimensionless factors is 1 J/kg. It is named after the Swedish physicst Rolf Sievert (1896-1966).

■ significant figures

measurements are not infinitely accurate: we must estimate measurement uncertainty. The number of significant figures is all of the certain digits plus the first uncertain digit.

Rules for significant figures:

1. Disregard all initial zeros.
2. Disregard all final zeros unless they follow a decimal point.
3. All remaining digits including zeros between nonzero digits are significant.

0.0023

have two significant figures

0.109

have three significant figures

2.00

have three significant figures

70

have one significant figure

In adition and subtraction, the number of significant figures in the answer depends on the original number in the calculation that has the fewest digits to the right of the decimal point.

In multiplication and division, the number of significant figures in a calculated result is determined by the original measurement that has the fewest number of significant digits.

In a logarithm of a number, keep as many digits to the right of the decimal point as there are significant figures in the original number.

In an antilogarithm of a number, keep as many digits as there are digits to the right of the decimal point in the original number.

silanes

silanes are saturated silicon hydrides, analogues of the alkanes (compounds of the general formula Si_nH_{2n+2}).

silanol

the Si-OH group found on the surface of silica gel. There are different strengths of silanols, depending on their location and relationship to each other. The strongest silanols are acidic and often lead to undesirable interactions with basic compounds during chromatography.

silica gel

the most commonly u.sed packing in liquid chromatography. It has an amorphous structure, is porous, and consists of siloxane and silanol groups. It is used as a bare packing for adsorption, as the suppon in liquid-liquid chromatography or for chemically bonded phases, and. with various pore sizes. as packing in size-exclusion chromatography. Microparticulate silicas of 5- and 10-mym average particle diameter are used in HPLC.

silicate

any one of a large family of substances chiefly used with titanium dioxide, the primary pigment, as an extender pigment. When used in moderation, these silicates (magnesium silicate, aluminum silicate, etc.) are valuable in helping control gloss, aid brushability, and increase hold-out properties and overall exterior durability.

silicone

a widely used group of oils and compounds derived from the mineral silica. Silicones are water repellent and very stable. There is no known toxicity when used externally. Silicones are polymeric or oligomeric siloxanes, usually considered unbranched, of general formula $(\text{-OSiR}_2\text{-})_n$ (R not equal to H).

■ siloxanes

siloxanes are saturated silicon-oxygen hydrides with unbranched or branched chains of alternating silicon and oxygen atoms (-Si-O-Si-).

■ silylation

the reaction of an organochloro-or orga noalkoxysilane with a compound containing a reactive group. In liquid chromatography, it refers to the process of derivatising the solute before chromatography in order to make it detectable or to prevent unwanted stationary phase interactions. It can also refer to the process of adding a chemically bonded phase to a solid support or to deactivating the packing to cut down on surface activity.

■ simethicone

a mixture of dimethicone and silica, used as a conditioning agent. It has no known toxicity.

■ SIMM

single in-line memory module.

■ single-strand chain

a chain that comprises constitutional units connected in such a way that adjacent constitutional units are joined to each other through two atoms, one on each constitutional unit.

■ single-strand macromolecule

a macromolecule that comprises constitutional units connected in such a way that adjacent constitutional units are joined to each other through two atoms, one on each constitutional unit.

■ single-strand polymer

a polymer , the macromolecules of which are singlestrand macromolecules.

■ sinks

ecosystems which absorb carbon dioxide naturally through micro-organisms and plants (i.e. forests and oceans).

■ sintering

the process in which fine particles of a material become chemically bonded at a temperature that is sufficient for atomic diffusion.

■ SIT

spontaneous ignition temperature. The temperature at which an oil ignites of its own accord in the presence of air or oxygen under standard conditions.

■ site characterisation

activities, whether in the laboratory or in the field, undertaken to establish the geologic and hydrologic conditions and the ranges of the parameters of a candidate site relevant to the location of a repository. These activities include borings, surface excavations, subsurface excavations and borings, and in situ testing needed to evaluate the suitability of a candidate site for the location of a repository but do not include preliminary borings and geophysical testing needed to assess whether site characterisation should be undertaken.

■ size-exclusion chromatography

same as steric exclusion chromatography.

■ skeletal atom

an atom in a skeletal structure.

■ skeletal bond

a bond connecting two skeletal atoms.

■ skeletal structure

the sequence of atoms in the constitutional unit(s) of a macromolecule, an oligomer molecule, a block, or a chain, which defines the essential topological representation.

■ slagging

formation of hard deposits on boiler tubes and/or piston crowns, usually due to the presence of sodium, vanadium and sulphur.

■ sludge

deposits in fuel tanks and caused by the presence of wax, sand, scale, asphaltenes, tars, water, etc.

■ slurry packing

the technique most ohen used to pack HPLC columns with microparticles. The packing is suspended in a slurry (10% wt/vol) and is rapidly pumped into the empty column. Special highpressure pumps are used.

■ smog

smog is a mixture of smoke and fog. The term is used to describe city fogs in which there is a large proportion of particulate matter (tiny pieces of carbon from exhausts) and also a high concentration of sulphur and nitrogen gases.

■ soap chromatography

an early name for ion-pair chromatography. Long-chain soaps or detergents were used as mobile phase additives.

■ sodium ascorbate

salt form of vitamin C used as an antioxidant and preservative in cosmetics.

■ sodium benzoate

an antiseptic and preservative used in food and cosmetics. It is non-toxic externally. A food grade preservative -sodium salt of benzoic acid. Can be found in nature in free and combined form e.g. cranberries, prunes and cinnamon.

■ sodium bicarbonate

an alkaline substance used in deodorants and toothpastes for its absorption properties and as a pH neutraliser.

■ sodium borate

see **borax**.

■ sodium carbonate

known as washing soda when hydrated. It is used in glass, soap and chemical detergents. Sensitisation is possible.

sodium laureth sulphate

similar to sodium lauryl sulphate except that it contains ethylene oxide to render it more water-soluble and consequently milder.

sodium lauryl sulphate

a synthetic detergent of which the origins may be plant (coconut and palm kernel oil) or animal. It is one of the most common ingredients used in shampoos and toothpastes for its foaming and detergent properties. It can be irritant in concentration, but correct formulation renders them functional with low irritancy. However, sensitive individual have reported an irritated scalp after using shampoo containing SLS. It has been found to cause damage to the delicate eye tissues - causing cataracts, retarding healing and preventing children's eyes from developing properly. Other research has shown that SLS and SLES may cause potentially carcinogenic nitrates and dioxins to form in shampoos and cleansers by reacting with commonly used ingredients found in many products.

sol

sols are dispersions of small solid particles in a liquid. The particles may be macromolecules or may be clusters of small molecules. Lyophobic sols are those in which there is no affinity between the dispersed phase and the liquid (e.g. silver chloride dispersed in water). Lyophobic sols are inherently unstable, in time the particles aggregate, and form a precipitate. Lyiophilic sols, on the other hand, are more like true solutions in which the solute molecules are large and have an affinity for the solvent (e.g. starch in water). Association colloids are systems in which the dispersed phase consists of clusters of molecules that have lyophobic and lyophilic parts (e.g. soap in water).

solar irradiance

energy emitted by the Sun.

sol-gel coating

a coating produced by the sol-gel process of glassmaking, in which glass is formed at low temperatures from suitable compounds by chemical polymerisation in a liquid phase; a gel is formed from which glass may be derived by the successive elimination of interstitial liquid and the collapse of the resulting solid residue by sintering.

solid form of matter

a solid holds its shape and volume even when not in a container. The molecules of a solid are tightly compacted and move only slightly.

solid support

same as support.

Solid-Phase Extraction (SPE)

a sample-preparation technique that uses a solid-phase packing contained in a small plastic cartridge. The solid stationary phases are the same as HPLC packings; however. the principle is different from HPLC. Sometimes referred to as digital chromatography. The process as most ohen practiced requires four steps conditioning the sorbent, adding the sample, washing away the impurities, and eluting the sample in as small a volume as possible with a strong solvent.

■ solubility

solubility is the maximum amount of solute that dissolves in a given quantity of solvent at a specific temperature. Generally, for a solid in a liquid, solubility increases with temperature; for a gas, solubility decreases. Common measures of solubility include the mass of solute per unit mass of solution (mass fraction), mole fraction of solute, molality, molarity, and others.

■ solubility product constant

solubility product constant (K_{sp}) (or the solubility product) is the product of the molar concentrations of the constituent ions, each raised to the power of its stoichiometric coefficient in the equilibrium equation. For instance, if a compound A_aB_b is in equilibrium with its solution

$A_aB_b(s)$ à $aA^+ + bB$

the solubility product is given by

$K_{sp}(A_aB_b) = [A^+]^a \cdot [B^-]^b$

■ soluble

matter or compounds capable of dissolving into a solution.

■ solute

the dissolved component of a mixture th tt is to be separated in the chromatographic column.

■ solvent

a solvent is a substance capable of reataining its physical state whilst forming a homogenous mixture with one or more substances. Some common solvents are listed below:

Solvent	Bpt °C
Water	100
Methanol	64
Ethanol	78
Acetone	56
Chlorofrom	61
Carbon tetrachloride	77
Benzene	80
Toluene	111
Tetrahydrofuran	66

■ solvent extraction

1. the removal of soluble material from a solid mixture by means of a solvent or the removal of components from a liquid mixture by use of a solvent with which the liquid is unmixable, or nearly so.
2. a traditional refinery process that is used to upgrade chemical and physical properties in the manufacture of lube oil base stocks. The process relies on the solubility of impurities (especially aromatic components that may also contain sulfur and nitrogen) in an extractive solvent, usually phenol. The by-products of this process is highly aromatic extract used to make Extender oils, and as feed for other refinery processes.

■ solvent strength

refers to the ability of a solvent to elute a particular solute or compound from a column. Described by Lloyd Snyder for LEAC (LSC) adsorption chromatography on alumina solvents were quantitatively rated in an eluotropic series. No eluotropic series exists for other modes.

■ solvents

solvent is a very general term, which can be applied to any substance that is used to carry another chemical. Water can be a solvent. Many solvents are completely or almost completely harmless to the environment. Many however, are very damaging, especially chlorinated solvents such as 1,1,1 Trichloroethane. These types of solvents are used to carry fluorocarbon treatments onto a fabric.

■ sorbent

refers to an adsorption packing used in liquid chromatography. A common sorbent is silica gel.

■ sorption

the binding, on a microscopic scale, of one substance to another. A term that includes both adsorption and absorption. The sorption of dissolved radionuclides onto aquifer solids or waste package materials by means of close-range chemical or physical forces is an important process modeled in this study. Sorption is a function of the chemistry of the radioisotopes, the fluid in which they are carried, and the mineral material they encounter along the flow path.

■ sorption coefficient (k_d)

a factor to calculate sorption of one substance to another (e.g., sorption of a radionuclide to a colloid or sorption of a radionuclide to the rock).

■ source term

types and amounts of radionuclides that are the source of a potential release of radioactivity from the repository.

■ specific gravity

specific gravity is ratio of the density of a material to that density of a water. Since one must specify the temperature of both the sample and the water to have a precisely defined quantity, the use of this term is now discouraged.

■ specific heat

specific heat is the quantity of heat required to raise the temperature of one gram of a substance by one degree Celsius.

■ specific quantity

specific quantity is often convenient to express an extensive quantity (e.g., volume, enthalpy, heat capacity, etc.) as the actual value divided by mass. The resulting quantity is called specific volume, specific enthalpy, etc.

■ specific surface area

the surface area of an LC packing based on measurement by an accepted technique, such as the BET method using nitrogen adsorption .

■ **specifications**

term referring to the properties of a given crude oil or petroleum product, which are 'specified' since they often vary widely even within the same grade of product. In the normal process of negotiation, seller will guarantee buyer that product or crude to be sold will meet certain specified limits, and will agree to have such limits certified in writing.

Generally the major qualities of oil for which a buyer would demand a guarantee are: API gravity (or specific gravity, in some cases), sulphur percentage measured by weight, pour point measured by degrees C maximum, viscosity min./max., BS&W percentage by weight, etc.

■ **spectroscopy**

the study of the interaction between electromagnetic radiation (light) and matter.

■ **spectrum (*plural is* spectra)**

a plot of intensity of light (either absorbed or emitted) versus energy.

■ **spent nuclear fuel**

fuel and the associated hardware withdrawn from a nuclear reactor following irradiation, the constituent elements of which have not been separated by reprocessing. Spent fuel that has been burned (irradiated) in a reactor to the extent that it no longer makes an efficient contribution to a nuclear chain reaction. This fuel is more radioactive than it was before irradiation, and it is thermally hot.

■ **SPF**

sun protection factor, an indicator of the degree of sun protection that a sunscreen offers. As a rule, for each SPF number 20 minutes of sun protection is offered by correct use of the product. Low SPF products tend to be in the SPF range of 1-6; medium SPF 7-11; and high 12 to 30+. There is some debate as to whether an SPF over 20 does in fact offer greater protection. The SPF rating applies to UVB protection only.

■ **spherical packing**

refers to spherical solid packing materials. Spherical packings are generally preferred over irregular particles.

■ **spindle**

contains fibers which form a network of achromatic filaments that extend inward from the poles of a dividing cell, forming a spindle-shaped figure.

They move the centromeres around, causing the rest of each chromosome to follow. This process ensures that each chromosome moves to its proper place during mitosis, when a cell divides to give rise to two cells, and during meiosis, the process of cell division that gives rise to eggs or sperm.

■ **spiro chain**

a double-strand chain consisting of an uninterrupted sequence of rings, with adjacent rings having only one atom in common. A spiro chain is a double-strand chain with adjacent constitutional units joined to each other through three atoms, two on one side and one on the other side of each constitutional unit.

spiro macromolecule

a double-strand macromolecule consisting of an uninterrupted sequence of rings, with adjacent rings having only one atom in common. A spiro macromolecule is a double-strand macromolecule with adjacent constitutional units joined to each other through three atoms, two on one side and one on the other side of each constitutional unit.

splay

a branch of a fault or fault zone.

split

the two halves of an emplacement, performance confirmation, or reserve drift, separated by an exhaust raise.

spray-drying

a ceramic processing used to create a uniform die fill for automated pressing; the slurry is fed into the spray drier, where circulating hot air evapourates the water; the resulting powder is collected for further processing.

squalene

a component of various animal and vegetable oils, commercially extracted from shark-liver oil, which has lubricant and fixative properties. It has no known toxicity.

sr1

seconds redwood # 1 @ 100 F.

ssf

seconds saybolt furol.

ssu

seconds saybolt universal.

stabilise

to convert to a form that resists change. Organic material is stabilised by bacteria which convert the material to gases and other relatively inert substances. Stabilised organic material generally will not give off obnoxious odors.

standard deviation

1. for a set of observations or a frequency distribution, the square root of the average of the squared deviations from the mean divided by n-1 (where n is the sample size)
2. the square root of the variance.

standard electrode potential

standard electrode potential ($E°$) (standard reduction potentialis) are defined by measuring the potential relative to a standard hydrogen electrode using 1 mol solution at 25 °C. The convention is to designate the cell so that the oxidised form is written first. For example,

$Pt(s) | H_2(g)H^+(aq) | Zn^{2+}(aq) | Zn(s)$

The e.m.f. of this cell is -0.76 V and the standard electrode potential of the $Zn^{2+} | Zn$ half cell is -0.76 V.

standard hydrogen electrode

standard hydrogen electrode is a system in which hydrogen ion and gaseous hydrogen are present in their standard states. The convention is to designate the cell so that the standard hydrogen electrode is written first.

$H_2(g)$ à$2H^+(aq) + 2e^-$

The electrode is used as a reference (of zero) for the values of other standard electrode potentials.

standards

standards are materials containing a known concentration of an analyte. They provide a reference to determine unknown concentrations or to calibrate analytical instruments.

The accuracy of an analytical measurement is how close a result comes to the true value. Determining the accuracy of a measurement usually requires calibration of the analytical method with a known standard. This is often done with standards of several concentrations to make a calibration or working curve.

A primary standard is a reagent that is extremely pure, stable, has no waters of hydration, and has a high molecular weight.

A secondary standard is a standard that is prepared in the laboratory for a specific analysis. It is usually standardised against a primary standard.

star macromolecule

a macromolecule containing a single branch point from which linear chains (arms) emanate. A star macromolecule with *n* linear chains (arms) attached to the branch point is termed an *n*-star macromolecule, e.g., five-star macromolecule. If the arms of a star macromolecule are identical with respect to constitution and degree of polymerisation, the macromolecule is termed a regular star macromolecule. If different arms of a star macromolecule are composed of different monomeric units, the macromolecule is termed a variegated star macromolecule.

star polymer

a polymer composed of star macromolecules.

starch

a sugar polymer derived from plant sources, used in cosmetics for its absorbent and soothing properties. Sensitisation to various forms of starch is possible.

states of matter

a substance can be solid, liquid or gaseous. Substances can change between states, normally when heated or cooled to increase or decrease the energy of the particles.

static mixer

a motionless mixer which has a series of fixed, geometric elements enclosed within a tubular housing. The internal elements impart flow division and radial mixing to the media flowing through the housing to produce a uniform dilution of the production.

■ **stationary phase**

the immobile phase involved in the chromatographic process. The stationary phase in liquid chromatography can be a solid, a bonded or coated phase on a solid Suppon, or a wall-coated phase. The stationary phase used ohen characterises the LC mode. For example. silica gel is used in adsorption chromatography, an octadecylsilane bonded phase in reversed-phase chromatography, etc.

■ **statistical copolymer**

a copolymer consisting of macromolecules in which the sequential distribution of the monomeric units obeys known statistical laws.

■ **statistical copolymerisation**

a copolymerisation in which a statistical copolymer is formed.

■ **steady flow**

when the velocity of fluid at each location is constant, the velocity field is invariant with time and the flow is said to be steady.

■ **steady, non-unifrom flow**

in this kind of flow, conditions change from point to point but do not with time. e.g., Flow of a liquid at constant flow rate through a tapering pipe running completely full.

■ **steady, uniform flow**

in this kind of flow, conditions do not change with position or time in the fluid e.g., Flow of liquid through a pipe of uniform bore running completely full at constant velocity.

■ **steady-state modelling**

modelling a system under the assumption that the variables are not changing with time. For example, flow fields can be simulated at a steady state if the boundary conditions, saturations, and fluxes are not changing with time.

■ **stearalkonium chloride**

see **quaternary ammonium compound**.

■ **steareth compounds**

compounds produced by ethoxylation of stearyl alcohol, which are used as solubilisers and co-emulsifiers.

■ **stearic acid**

a waxy, fatty acid that can be derived from tallow and other animal fats, but also from cocoa butter and other vegetable fats. It is used as an emulsiiier and emollient in soaps, creams and lotions. It is non-toxic. Sensitisation is unusual but possible.

■ **stepwise elution**

use of eluents of different compositions during the chromatographic run. These eluent~ are added in a stepwise manner with a pump, or by a selector valve. Gradient elution incorporates continuous changing of solvent composition.

■ **stereoblock macromolecule**

a block macromolecule composed of stereoregular, and possibly nonstereoregular, blocks.

■ **stereoblock polymer**

a polymer composed of stereoblock macromolecules.

■ **stereoregular macromolecule**

a regular macromolecule essentially comprising only one species of stereorepeating unit.

■ **stereoregular polymer**

a polymer composed of stereoregular macromolecules.

■ **stereorepeating unit**

a configurational repeating unit having defined configuration at all sites of stereoisomerism in the main chain of a regular macromolecule, a regular oligomer molecule, a regular block, or a regular chain.

■ **steric exclusion chromatography**

a major LC mode in which samples are separated by virtue of their size in solution. Also known as size-exclusion. gel permeation, gel filtration, or gel chromatography.

■ **stevia rebaudiana**

stevia - a herbal flavouring - beneficial for the health of the gums.

■ **stimulant**

herb that increases the activity or efficiency of a system or organ; acts more rapidly than a tonic. Examples: cayenne, camphor, ephedra, barberry extract, yellow thistle juice, sandalwood, gotu kola, guggul, myrrh.

■ **stochastic**

involving a variable (e.g., temperature, porosity) that may take on values of a specified set with a certain probability. Data from a stochastic process is an ordered set of observations, each of which is one item from a probability distribution.

■ **stoichiometric number**

stoichiometric number (*t*) is the number appearing before the symbol for each compound in the equation for a chemical reaction. By convention, it is negative for reactants and positive for products.

■ **stoichiometry**

stoichiometry is the relative proportions in which elements from compounds or in which substances react. Every chemical reaction has its characteristic proportions. For example, when methane unites with oxygen in complete combustion, 1 mol of methane requires 2 mol of oxygen.

$CH_4 + 2O_2$ à $CO_2 + 2H_2O$

At the same time, 1 mol of carbon dioxide and 2 mol of water are formed as reaction productions.

Alternatively, 16 g of methane and 64 g of oxygen produce 44 g of carbon dioxide and 36 g of water.

■ **stoke**

the unit of kinematic viscosity.

■ **stokesian**

a stokesian (or non-Newtonian) fluid is a viscous fluid whose shear stresses are a non-linear function of the fluid strain rate.

■ **STP**

standard temperature (273.15 K) and pressure (101 325 Pa) is the standard conditions used as a basis for calculations involving quantities that vary with temperature and pressure. These conditions are used when comparing the properties of gases.

■ **straight-phase chromatography**

same as Normalphase chromatography.

■ **straight-run**

refers to a petroleum product produced by the primary distillation of crude oil, free of cracked components.

■ **stratification**

occurs in blended fuels that have a compatibility problem. It is usually experienced when paraffinic based oils are mixed with asphaltic based oils, causing asphaltenes to precipitate and settle to the bottom of the tank.

■ **stratosphere**

stratosphere is the part of the earth's atmosphere extending from the top of the troposphere (typically 10 km to 15 km above the surface) to about 50 km. It is characterised by an increase in temperature with increasing altitude.

■ **streamline**

a path in a steady flow field along which a given fluid particle travels.

■ **stress corrosion cracking**

preferential corrosion initiation in response to unrelieved tensile stresses in an aggressive corrosion environment.

■ **stress intensity**

the amount of stress at a given point in a structure. Derived from combined totals of both positive (tension) stress and negative (compression) stress.

■ **subchain**

an arbitrarily chosen contiguous sequence of constitutional units in a chain.

■ **substrate**

a reactant molecule that is capable of binding to the active site of an enzyme.

A substrate is a reactant molecule that is capable of binding to the active site of an enzyme. When the substrate binds to the enzyme the specific reaction is carried out at a much faster rate than without the binding.

■ **subsurface facilities**

the repository's underground structures and systems. The subsurface facilities include the main drifts, exhaust mains, turnouts, emplacement drifts, ventilation shafts, mechanical and structural support systems, underground utilities, waste emplacement and retrieval equipment, and surface-based control systems.

■ **sulphate aerosols**

particulate matter that consists of compounds of sulphur formed by the interaction of sulphur dioxide and sulphur trioxide with other compounds in the atmosphere. Sulphate aerosols are injected into the atmosphere from the combustion of fossil fuels and the eruption of volcanoes like Mt. Pinatubo.

■ **sulphur**

an element that is present in crude oil and natural gas as an impurity in the form of its various compounds.

■ **sunscreens**

substances used to block out the sun's burning rays. Sunscreens may act in two main ways: chemical sunscreens, for example PABA, which retract the sun's rays; and physical sunscreens, for example titanium dioxide, which block the sun's rays from reaching tile skin. Sunscreens are also divided into those that protect against UVB light, which causes immediate burning (and this is what the SPF system is based on); and those that protect against UVA light, which is thought to cause longer term but less immediately obvious damage and premature ageing.

■ **superconductor**

superconductor is a material that experiences a nearly total loss of electrical resistivity below a critical temperature T_c.

■ **supercritical**

a condition in which a substance experiences a change in its pressure, temperature, or volume that would normally carry the system through its critical point although no phase change is observed.

■ **supercritical fluid chromatography**

a technique that uses a supercritical fluid as the mobile phase. The technique has been applied to the separation of substances that cannot be handled effectively by liquid chromatography (because of detection problems) or gas chromatography (because of the lack of volatility). Examples are separations of triglycerides, hydrocarbons, and fatty acids. GC detectors and HPLC pumps have been used together in SFC.

■ **superficially porous packing**

see Porous-layer bead.

■ **superfluid**

superfluid is a fluid with near-zero viscosity and extremely high thermal conductivity. Liquid helium exhibits these properties below 2.186 K (the ë point).

■ **support**

refers to solid particles. Suppon can be naked or coated or can have a chemically bonded phase in HPLC.

■ **supporting ligands**

the products of the bonding reactions that deposit gas molecules on a substrate in a CVD chamber.

■ **suppressor column**

in ion chromatography. refers to the column placed after the ion-exchange column Its purpose is to remove or suppress the ionisation of buffer ions so that sample ions can be observed in a weakly conducting background with a conductivity detector.

surface area

in an adsorbent, refers to the total area of the solid surface as determined by an accepted measurement technique such as the BET method using nitrogen adsorption. The surface area of a typical porous adsorbent such as silica gel can vary from 100 to 600 m^2/g.

surface coverage

usually refers to the mass of stationary phase per unit area of an LC suppon. Ohen expressed in mymol/m~ of surface. Sometimes the %C is given as an indicator of surface coverage.

surface facilities

all permanent facilities within the restricted area constructed in support of site characterisation activities and repository construction, operation, and closure activities, including surface structures, utility lines, roads, railroads, and similar facilities, but excluding the underground facility.

surface tension

the surface tension of a liquid is the work that must be done to bring enough molecules from inside the liquid to the surface to form one new unit area of that surface ($J/m^2 = N/m$). Historically surface tensions have been reported in handbooks in dynes per centimeter (1 dyn/cm = 0.001 N/m).

Surface tension is the tendency of the surface of a liquid to behave like a stretched elastic membrane. There is a natural tendency for liquids to minimise their surface area. For this reason, drops of liquid tend to take a spherical shape in order to minimise surface area. For such a small droplet, surface tension will cause an increase of internal pressure p in order to balance the surface force. This is the attractive forces which liquid molecules have for each other.

surfactant

a substance with the ability to reduce the surface tension at the interface between two unlike surfaces. Soap is an example, as are all detergents. Amphoteric surfactants have both a positive (cationic) group and a negative (anionic) group. The final pH of a product will dictate which group is more dominant. At an intermediate pH, both forms are present. At a pH of less than 7 the cationic group is more prevalent, and at a pH of over 7 the anionic group is more prevalent. They tend to be very mild and are often found in baby products. Anionic surfactants are negatively charged surface active agents used widely in shampoo and foaming baths as the primary cleansing agent. These are chemically produced, although the raw materials used to produce them can be vegetable, animal or mineral derived. Examples include ammonium lauryl sulphate, sodium laureth sulphate. Cationic surfactants are detergents of which the ions are positively charged. They are used mainly in conditioners.

■ **suspension**

the process of a cleaning agent holding insoluble dirt particles in the cleaning solution and keeping them from redepositing on a clean floor.

■ **swelling**

process in which resins and gels increase their volume because of their solvent environment. Solvent enters ion-exchange resin to dilute ions in gels, solvent penetrates pores. If swelling occurs in packed columns. blockage or increased back pressure can occur. In addition. column efficiency can be affected.

■ **synapses**

the gaps between neurons. They allow messages in the form of electrical currents to be passes between the neurons, allowing thought , sense, and action.

■ **syndiotactic macromolecule**

a tactic macromolecule , essentially comprising alternating enantiomeric configurational base units , which have chiral or prochiral atoms in the main chain in a unique arrangement with respect to their adjacent constitutional units. In a syndiotactic macromolecule, the configurational repeating unit consists of two configurational base units that are enantiomeric. A syndiotactic macromolecule consists of racemo diads.

■ **syndiotactic polymer**

a polymer composed of syndiotactic macromolecules.

■ **synergistic**

chemicals that when combined have a greater effect than the sum of the two independently.

■ **synthetic detergents**

these are sometimes called soapless detergents. They are typically made from by-products of refining crude oil. They do not form a scum in hard water and lather better than soaps.

■ **synthetic lubricants**

lube oils possessing a base oil that has been manufactured from chemical constituents or by the polymerisation of hydrocarbons (olefins) rather than by conventional refining of petroleum. The three most common types of synthetic base oils are:

Polyalpolefins

Organic esters

Polyglycols

Synthetic lubricants have several advantages over conventional mineral oils:

excellent low temperature fluidity

low pour point

high natural viscosity index

excellent oxidation stability

high flash, fire, and auto-ignition points

low volatility

non-corrosive and non-toxic

Synthetic lubricants have been in use for some time in applications such as jet engine lubrication, Arctic lubrication, and fire-resistant hydraulic fluids. These applications tolerate the extremely high cost of synthetics because they are the only products that do the job. Synthetic lubricants are now beginning to replace conventional petroleum lubricants in some applications. Despite their higher purchase price, synthetics may offer operating advantages that can make them more economical in the long run. (For example, reduced oil consumption, longer oil life, improved fuel economy, and easier starting at low temperatures).

■ system model

the analytical tool to examine the future behaviour of the potential repository and its component barriers.

■ system performance

the complete behaviour of a geologic repository system at Yucca Mountain in response to the features, events, and processes that may affect it.

■ tactic macromolecule

a regular macromolecule in which essentially all the configurational (repeating) units are identical.

■ tactic polymer

a polymer composed of tactic macromolecules.

■ tacticity

the orderliness of the succession of configurational repeating units in the main chain of a regular macromolecule, a regular oligomer molecule, a regular block, or a regular chain.

■ Tag-Robinson colourimeter

an instrument used to determine the colour of oils. Also a scale of colour values.

■ tailing

the phenomenon in which the normal Gaussian peak has an asymmetry factor > 1. The peak wil I have skew in trailing edge. Tailing is caused by sites on the packing that have a stronger-than normal retention for the solute. A typical example of a tailing phenomenon is the strong adsorption of amines on the residual silanol groups of a low-coverage reversed-phase packing.

■ TBN

Total Base Number. This is measured in mg. KOH needed to neutralise an acidic solution through a reverse titration. TBN is the ability of the product to neutralise acid. In a motor oil, this is a property which allows the oil to neutralise acids from combustion that would otherwise degrade the oil.

■ technical ceramic

a ceramic that exhibits a high degree of industrial efficiency through carefully designed microstructures and superb dimensional precision.

■ tectonic

pertaining to geologic forms or effects created by deformation of the earth's crust.

■ teflon

teflon is a polymer which is widely used in safety clothing because of its excellent resistance to chemicals, and good thermal resistance. However, it has poor mechanical properties, so is generally combined with other materials to provide the required mechanical durability.

■ **temperature**

temperature is measure to the average kinetic energy of its molecules. The SI unit in which thermodynamic temperature is expressed is the kelvin (K).

■ **temperature of no return**

the temperature at which the rate of heat generation is equal to the maximum rate of cooling available.

■ **temperature programming**

changing column temperature as a function of time during the separation. Rarely used in HPLC if so, usually in a stepwise manner.

■ **tertiary**

the first of two geologic periods of the Cenozoic Era extending from the end of the Mesozoic Era to the beginning of the Quarternary Period, covering a time span from about 65 million to about 2 million years ago.

■ **tesla**

tesla (T) is the SI derived unit of magnetic flux density. The tesla is magnetic flux density of a magnetic flux of one weber per square metre (T = Wb/m^2). The unit is named after the Croatian scientist Nikola Tesla (1857-1943), equal to V. s/m^2.

■ **theoretical plate**

a concept described by Manin and Synge. Relates chromatographic separation to the theory of distillation. Measure of column efficiency. Length of column relating to this concept is called height equivalent to a theoretical plate (HETP). see **HETP.**

■ **theoretical yield**

theoretical yield is the maximum quantity of a product that could be formed in a chemical reaction if all the limiting reactant reacted to form products (distinguished from actual yield).

■ **thermal conductivity**

thermal conductivity (*Ë*) is rate of heat flow divided by area and by temperature gradient.

■ **thermal loading**

1. the areal density at which waste packages are emplaced within the repository as characterised by the areal power density and the areal mass loading.
2. the application of heat to a system, usually measured in terms of Watt density. The thermal loading for a repository is the result of heat produced by the radioactive waste in the active disposal area.

■ **thermal runaway**

see **runaway**.

■ **thermal stress**

stress caused by temperature changes in a material that is physically restricted and unable to expand or contract accordingly.

■ **thermal value**

calories per gram of BTU per pound produced by burning fuels.

■ **thermally unstable**

chemicals and materials are thermally unstable if they decompose, degrade or react as a function of temperature and time at or about the temperature of use.

■ **thermal-mechanical effects**

changes in the geomechanical properties of the repository host rock produced by heating of the rock associated with the emplacement of radioactive waste in the repository. An example might be decreased rock strength related to increased fracturing caused by heating of the rock.

■ **thermodynamic data**

data associated with the aspects of a reaction that are based on the thermodynamic laws of energy, such as Gibbs' free energy, and the enthalpy (heat) of reaction.

■ **thermodynamic laws**

thermodynamic laws are the foundation of the science of thermodynamics:

First law: The internal energy of an isolated system is constant; if energy is supplied to the system in the form of heat dq and work dw, then the change in energy $dU = dq + dw$.

Second law: No process is possible in which the only result is the transfer of heat from a reservoir and its complete conversion to work.

Third law: The entropy of a perfect crystal approaches zero as the thermodynamic temperature approaches zero.

■ **thermodynamics**

thermodynamics is the scientific study of the interconversion of heat and other forms of energy.

■ **thermophiles**

a class of organisms physically suited to survive and prosper in regions of unusually high or unusually low temperatures.

■ **thermosphere**

thermosphere is the layer of the earth's atmosphere extending from the top of the mesosphere (80 km - 90 km above the surface) to about 500 km. It is characterised by a rapid increase in temperature with increasing altitude up to about 200 km, followed by a leveling off in the 300 km - 500 km region.

■ **thin film**

any of several thin layers (generally less than 1micrometer each) of insulating, conducting, or semiconductor material that are deposited successively on a supporting substrate in precise patterns to collectively form all or part of an integrated circuit; the deposition can be performed by mechanical, chemical, or high-vacuum evapouration methods.

■ **thinner**

the thinner and binder together form the paint's vehicle. Water, the thinner used in latex paints, evaporates as the paint dries, allowing a smooth paint application. Turpentine or spirits are the thinners in oil-based paints.

■ thixotropic

adjective which describes full-bodied material which undergoes a reduction in viscosity when shaken, stirred, or otherwise mechanically disturbed and which readily recovers the full-bodied condition on standing.

■ thixotropic fluids

fluids for which the dynamic viscosity decreases with the time for which shearing forces are applied. e.g. thixotropic jelly paints.

■ three way safety bulb

the three way safety bulb is used for pipeting The attachment is placed over the mouth of the pipet. The air (A) valve is used to empty the bulb of air. The suction (S) valve is used to draw liquid into the pipet. The empty (E) valve is used to drain liquid out of a pipet.

■ three-dimensional model

a three-dimensional representation of physical conditions and/or processes.

■ time constant

same as lag time.

■ timken ok load

this is a measure of the extreme pressure properties of a lubricant. Lubricated by the product under investigation, a standard steel roller rotates against a block. Timken OK load is the heaviest load that can be carried without scoring

■ tincture

an extract (usuallyt herb) that has been prepared using alcohol and water to extract plant components.

■ titania

TiO_2, an uncommon adsorbent used in ad sorption chromatography.

■ titanium dioxide, anatase (TiO_2)

a high opacity, bright white pigment of the chalking type, used as a prime pigment in paints, rubber, plastics. Prepared from the mineral ilmenite, or rutile ore.

■ titanium dioxide, rutile TiO_2)

a high opacity, bright white pigment of the non-chalking type, used as a prime pigment in paints, rubber, plastics. Prepared from the mineral ilmenite, or rutile ore.

■ titration

a procedure that uses a neutralisation reaction to determine the normality (the number of equivalents per liter of solution) of an unknown acid or base solution.

■ titration curve

titration curve is a graphical representation of the amount of a species present (often hydrogen ion) vs. volume of solution added during a titration.

■ TMR

abbreviation for time to maximum rate. time taken for a material to self heat to the maximum rate of decomposition from a specific temperature, usually under adiabatic conditions.

■ tmrad

time to maximum rate from a specified temperature under adiabatic conditions.

■ tonne

tonne (t) is an alternative name for megagram (1000 kg).

■ topped crude oil

oil from which the light ends have been removed by a simple refining process. Also referred to as 'reduced crude oil'.

■ tortuosity

a propeny of a packed column that indicates the degree of unevenness of the path followed by the solute molecule as it passes down the column. Covered in the A-term of the van Deemter equation.

■ total effective dose equivalent

for purposes of assessing doses to workers, the sum of the deep-dose equivalent (for external exposures) and the committed effective dose equivalent (for internal exposures). For purposes of assessing doses to members of the public (including the reasonably maximally exposed individual), total effective dose equivalent means the sum of the effective dose equivalent (for external exposures) and the committed effective dose equivalent (for internal exposures).

■ total existent sediment

combination of inorganic and hydrocarbon sediments existing in a fuel as delivered.

■ total permeation volume (v_p)

the retention volume on an SEC packing in which all molecules smaller than the smallest pore will elute. In other words, at V_p, all molecules totally permeate all of the pores and elute together.

■ Total System Performance Assessment (TSPA)

a risk assessment that quantitatively estimates how the proposed Yucca Mountain disposal system will perform in the future under the influence of specific features, events, and processes, incorporating uncertainty in the models and data.

■ tough2

a computer software code used to simulate three-dimensional flow of groundwater and heat in unsaturated or saturated porous and fractured media. It is the basis for the unsaturated zone flow process model.

■ toughreact

thermal-hydrologic-chemical software code used to simulate the water composition on the drip shield and waste package.

■ toxic

a substance which is poisonous to a living organism.

■ **toxicity**

the relative degree of being poisonous or toxic. A condition which may exist in wastes and will inhibit or destroy the growth or function of certain organisms.

■ **trace enrichment**

technique in which trace amounts of compounds from a weak mobile phase or solution are retained on an HPLC packing and then are eluted by the addition of a stronger mobile phase in a concentrated form. The technique has been most successfully applied in the concentration of traceamounts of hydrophobic compounds (e.g. polynuclear aromatic hydrocarbons) out of water using a reversed-phase column. A strong Solvent such as acetonitrile serves to elute the enriched compounds.

■ **tracer**

a substance or dye used in hydrologic tests to observe the movement of groundwater and sorbing and nonsorbing chemical species. Also applicable to gas injection tests using gaseous tracers to measure the breakthoughs at observation points.

■ **tragacanth**

a gum which is used as a thickener. Sensitisation is possible.

■ **transistor**

a three terminal semiconductor device used for amplification, switching and detection, typically containing two rectifying junctions and characteristically operating so that the current between the other pair, one terminal being common to input and output is controlled by the third terminal.

■ **transition metals**

this group of metals is distinguished from other metals not by their physical properties, but by their electronic structure. Transition metals are elements characterised by a partially filled *d* subshell. The First Transition Series comprises scandium (Sc), titanium (Ti), vanadium (V), chromium (Cr), manganese (Mn), iron (Fe), cobalt (Co), nickel (Ni) and copper (Cu). The Second and Third Transition Series include the lanthanides and actinides, respectively. The transition metals are noted for their variability in oxidation state. Thus, manganese has two electrons in its outside shell and five electrons in the next shell down, and exhibits oxidation states of +1, +2, +3, +4, +5, +6, and +7.
They are also characterised by the fact that well into the series, going from left to right, the properties of succeeding metals do not differ greatly from preceding ones.

■ **transmissive fracture**

a fracture in rock through which groundwater could flow.

■ **transmittance**

transmittance (θ) is ratio of the radiant or luminous flux at a given wavelength that is transmitted to that of the incident radiation. Also called transmission factor.

■ **transmutation**

the conversion of one element into another by a process taking place in the nucleus.

■ **transpiration**

the process by which water absorbed by plants, usually through the roots, is evapourated into the atmosphere from the plants' surfaces. It is an important process for removal of water that has infiltrated below the zone where it could be removed by evapouration.

■ **transport**

a process in which substances carried in groundwater move through the subsurface by means of the physical mechanisms of convection, diffusion, and dispersion and the chemical mechanisms of sorption, leaching, precipitation, dissolution, and complexation. Types of transport include advective, diffusive, and colloidal transport.

■ **transportation cask**

a heavily shielded container that meets applicable regulatory requirements used to ship spent nuclear fuel or high-level radioactive waste.

■ **transverse dispersion**

the spreading of a solute in groundwater in directions perpendicular to the direction of the groundwater flow path.

■ **triple point**

triple point is the point in p,T space where the solid, liquid, and gas phases of a substance are in thermodynamic equilibrium.

■ **troposphere**

troposphere is the lowest part of the earth's atmosphere, extending to 110 km to 15 km above the surface. It is characterised by a decrease in temperature with increasing altitude. The exact height varies with latitude and season.

■ **tumorigenic**

tumorigenic is a description which can be applied to any material or phenomenon (a chemical, a radiochemical, radiation, etc.) capable of generating tumours. Toxicant is a comparatively rarely-used term which describes any material which is potentially toxic.

■ **turbulent**

a flow field that cannot be described with streamlines in the absolute sense. However, time-averaged streamlines can be defined to describe the average behaviour of the flow. In turbulent flow, the inertia stresses dominate over the viscous stresses, leading to small-scale chaotic behaviour in the fluid motion.

■ **two-dimensional model**

1. a two-dimensional slice through an entity, such as the earth's crust, usually in the horizontal and vertical directions, on which known features are placed and are used to predict likely features that may exist between points of known data.

2. mathematically, a model that represents physical conditions or processes; this mathematical model is composed of both hori-

zontal rows and vertical columns of grid cells arrayed in L-shaped configurations only one grid cell thick.

ubbehohde viscosimeter

a suspended level apparatus for accurately determining the viscosity of a liquid.

ullage

the amount which a tank or vessel lacks of being full.

ultraviolet light

electromagnetic radiation with a wavelength shorter than that of visible light.

uncertainty

a measure of how much a calculated or estimated value may vary from the unknown true value.

underground facility

the underground structure, backfill materials, if any; and openings that penetrate the underground structure.

unified atomic mass unit

unified atomic mass unit (u or m_u) is a unit of mass used in atomic, molecular, and nuclear science, defined as the mass of one atom of ^{12}C divided by 12. Its approximate value is

$u = (1.660\,565 \pm 0.000\,008\,6) \cdot 10^{-27}$ kg.

uniform flow

uniform flow occurs when the magnitude and direction of velocity do not change from point to point in the fluid.

Flow of liquids through long pipelines of constant diameter is uniform whether flow is steady or unsteady.

uniform polymer

monodisperse polymer. A polymer composed of molecules uniform with respect to relative molecular mass and constitution. A polymer comprising a mixture of linear and branched chains, all of uniform relative molecular mass, is not uniform. A copolymer comprising linear molecules of uniform relative molecular mass and uniform elemental composition but different sequential arrangements of the various types of monomeric units, is not uniform (e.g., a copolymer comprising molecules with a random arrangement as well as a block arrangement of monomeric units). A polymer uniform with respect only to *either* relative molecular mass *or* constitution may be termed uniform, provided a suitable qualifier is used (e.g., 'a polymer uniform with respect to relative molecular mass'). The adjectives *monodisperse* and *polydisperse* are deeply rooted in the literature, despite the former being non-descriptive and self-contradictory. They are in common usage and it is recognised that they will continue to be used for some time; nevertheless, more satisfactory terms are clearly desirable. After an extensive search for possible replacements, the terms *uniform* and *non-uniform* have been selected and they are now the preferred adjectives.

unit cell

unit cell is the smallest fragment of the structure of a solid that by repetition can generate the entire structure.

■ **universal solvent**

water is called the universal solvent because it dissolves both ionic compounds and polar molecular compounds. Water usually cannot dissolve nonpolar molecules.

■ **unsaturated compound**

an organic compound that contains carbon-carbon multiple bonds. Examples include ethene and propene.

■ **unsaturated hydrocarbons**

unsaturated hydrocarbons are organic compounds containing double (alkenes) or triple (alkynes) bonds in its molecules.

■ **unsaturated solution**

unsaturated solution is a solution that contains less than the maximum possible equilibrium concentration of a solute.

■ **unsaturated zone**

the zone of soil or rock below the ground surface and above the water table.

■ **unstable substance/material**

substance or material that decomposes either in the pure state or in the state as normally produced.

■ **unzipping**

the splitting of the cladding on a fuel rod.

■ **uptake**

intake by and exposure of the receptor to a contaminant.

■ **urban heat island effect**

a warming effect found in urban areas due to the abundance of buildings, streets, concrete and asphalt — all things with low albedo, and thus a higher absorptivity of solar radiation. This heat is then radiated back into the atmosphere, warming the area.

■ **urea**

a product of protein metabolism present in urine. It is used in cosmetics and toiletries as an antiseptic and preservative. It has no known toxicity externally.

■ **use-dilution**

the final concentration at which a product is used.

■ **vacancy**

vacancy is a missing atom or ion in a crystal lattice.

■ **vacancy chromatography**

technique in which a mobile phase additive causes a positive detector signal output. When a solute elutes from the column, it dilutes the signal and gives a negative peak (a vacancy). The technique has been most recently applied to single-column ion chromatography, in which mobile phases that absorb in the UV region, such as citrate and phthalate buffers, are used. When a nonab sorbing anion elutes, it dilutes the UV-absorbing background and causes a negative peak the detector output leads are usually reversed so that the chrmatogram looks normal.

vacuum drying

the removal of liquid from a solid material while in a vacuum system, to lower the temperature at which evaporation takes place and thus prevent heat damage to the material.

vacuum pressure

whenever atmospheric pressure is used as a reference, the possibility exists that the pressure thus measured can be either positive or negative. Negative gauge pressure are also termed as *vacuum pressures*. Hence, if a gauge tapped into a tank indicates a vacuum pressure of 31 kPa, this can also be stated as 70 kPa absolute, or -31 kPa gauge, assuming that the atmospheric pressure is 101 kPa absolute.

van deemter equation

an equation used to explain band broadening in chromatography. The equation represents the height equivalent of a theoretical plate and has three terms. The A-term is used to describe eddy diffusion, which allows for the different paths a solute may follow when passing over particles of different sizes. The B-term is for the contribution caused by molecular diffusion or longitudinal diffusion of the solute while passing through the eolumn. The C term is the contribution of mass transfer and allows for the finite rate of transfer of the solute between the stationary phase and mobile phase. h = A + B/v + Cv.

van der waals bond

a secondary interatomic bond between adjacent molecular dipoles, which may be permanent or induced.

van der waals equation

van der Waals' equation is an equation of state for fluids which takes the form:
where p is pressure, V is molar volume, T is temperature, R is the molar gas constant, and a and b are characteristic parameters of the substance which describe the effect of attractive and repulsive intermolecular forces.

$$p = \frac{nRT}{V-nb} - a\left(\frac{n}{V}\right)^2$$

van der waals' force

van derWaals' force is the weak attractive force between two molecules which arises from electric dipole interactions. It can lead to the formation of stable but weakly bound dimer molecules or clusters. They are named after the Dutch physicist Johannes van der Waals (1837-1923).

van genuchten's capillary-strength parameter

a parameter in a functional relationship between saturation (or water content) and potential (a measure of the suction due to capillary forces).

vanadium inhibitor

an organic and/or inorganic metal bearing chemical intended to chemically and/or physically combine with the compounds formed during combustion of heavy fuel oil to improve the surface properties of the treated ash compounds.

■ van't Hoff equation

van't Hoff equation is the equation expressing the temperature dependence of the equilibrium constant *K* of a chemical reaction:

■ vapour pressure

1. describes a measure of a liquids tendency to evapourate. Every liquid has a characteristic vapour pressure that changes as the internal temperature of the liquid changes. Generally, as the temperature of a liquid increases, its vapour pressure also increases.

2. the pressure at which a liquid will boil is called its vapour pressure. This pressure is a function of temperature (vapour pressure increases with temperature). In this context we usually think about the temperature at which boiling occurs. For example, water boils at 100°C at sea-level atmospheric pressure (1 atm abs). However, in terms of vapour pressure, we can say that by increasing the temperature of water at sea level to 100 °C, we increase the vapour pressure to the point at which it is equal to the atmospheric pressure (1 atm abs), so that boiling occurs. It is easy to visualise that boiling can also occur in water at temperatures much below 100°C if the pressure in the water is reduced to its vapour pressure. For example, the vapour pressure of water at 10°C is 0.01 atm. Therefore, if the pressure within water at that temperature is reduced to that value, the water boils. Such boiling often occurs in flowing liquids, such as on the suction side of a pump. When such boiling does occur in the flowing liquids, vapour bubbles start growing in local regions of very low pressre and then collapse in regions of high downstream pressure. This phenomenon is called as **cavitation.**

■ variability (statistical)

a measure of how a quantity varies over time or space.

■ vehicle

the liquid portion of the paint, in which the pigment is dispersed; it is composed of a binder and a thinner.

■ velocity (u)

same as Linear velocity.

■ venting

emergency flow of vessel contents out of the vessel. The pressure is reduced by venting, thus, avoiding a vessel rupture due to overpressurisation. The vent flow can be single or multi-phase with consequent differences in mass flux and depressurisation characteristics.

■ vinegar

dilute acetic acid, used as an astringent and pH adjuster.

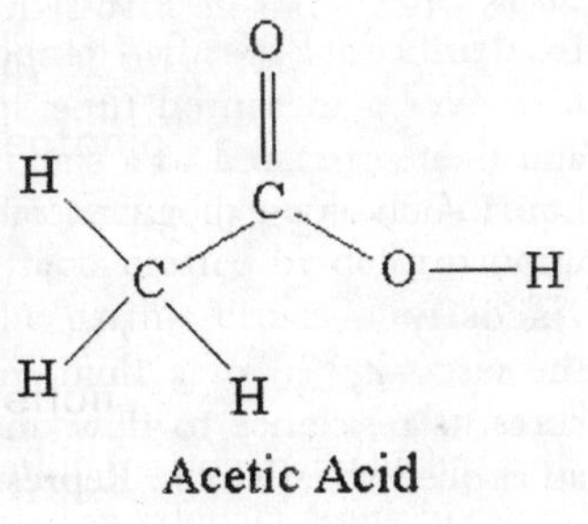

Acetic Acid

■ **vinyl**

the unsaturated, univalent radical CH_2: CH — derived from ethylene. Any of the various compounds containing this group, typically highly reactive, easily polymerised and used as a basic material for coatings and plastics. Any of the various plastics, typically tough and flexible.

■ **vis-breaking**

a light thermal cracking process carried out on a fuel oil during the refining process to reduce product viscosity without blending.

■ **visco-elastic fluids**

some fluids have elastic properties, which allow them to spring back when a shear force is released. e.g. egg white.

■ **viscoelasticity**

a type of deformation exhibiting the mechanical characteristics of viscous flow and elastic deformation.

■ **viscosimeter**

a device for determining the viscosity of oil. There are several methods or devices in general use. Basically, a fixed quantity of oil is allowed to pass through a fixed orifice at a specified temperature over a measured time span and then compared to a standard liquid such as a calibration oil or water.

viscosity

the viscosity (h) of a fluid measures its resistance to flow under an applied shear stress. Representative units for viscosity are kg/(m.sec), g/(cm.sec) (also known as poise designated by P). The centipoise (cP), one hundredth of a poise, is also a convenient unit, since the viscosity of water at room temperature is approximately 1 centipoise.

■ **viscosity index (v.i)**

the measure of the rate of change of viscosity with temperature. This change is common to all fluids - some more, some less. heating tends to make them thinner - cooling thicker. The higher the V.I., the less the tendency for the viscosity to change. V.I. is determined by formula from the viscosities at 40°C and 100°C. The latter test is required for V.I.'s above 100. High V.I. oils are often preferred for service in which a relatively constant viscosity is desired under conditions of varying temperature. Some hydraulic systems require this property. Paraffinic oils are inherently high in V.I. and the V.I. of any petroleum oil can be increased by the addition of a V.I. improper. Naphthenic oils are inherently low in V.I. and aromatic oils are still lower - often having negative numbers

■ **vitamins**

the name vitamins is obtained from 'vital amines' as it was originally thought that these substances were all amines. This is now known not to be as vitamins have a range of structures. The body requires a small amout of vitamins, but any deficiency leads to metabolic and physical disorders.

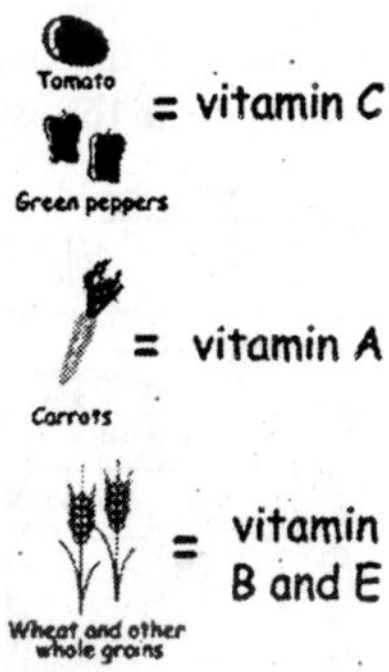

■ **viton**

viton is a hexafluoropropylene - vinylidene fluoride co-polymer which is widely used in protective clothing.

■ **vitric tuff**

volcanic rock composed of glassy shards of volcanic ash.

■ **vitrified high-level radioactive waste**

a type of processed high-level radioactive waste where the waste is mixed with glass-forming chemicals and put through a melting process. The melted mixture is then put into a canister where it becomes a dry, solid 'log' of waste in a glassy matrix.

■ **VOC (Volatile Organic Content)**

any carbon compound that evaporates under standard test conditions. Essentially, all paint solvents except water are VOCs.

■ **void**

the formation of a space. usually at the head of the column. Caused by a settling or dissolution of the packing. A void in the column leads to decreased efficiency and loss of resolution. Even a small void can be disastrous for small micropaniculate columns. The void can sometimes be removed by filling it with glass beads or porous packing

■ **void time (t_m or t_0)**

the time for elution of an unretained peak.

■ **void volume (v_i)**

the total volume of mobile phase in the column the remainder of the column is taken up by packing material. Can be determined by injecting an unretained substance that measures void volume plus extracolumn volume. Also referred to as interstitial volume. V_0 or V_m are sometimes used as symbols.

■ **volatile**

a volatile substance is one that is capable of being evaporated or changed to a vapor at a relatively low temperature. Volatile substances also can be partially removed by air stripping.

■ **volatility**

that property of a liquid that defines its evaporation characteristics. Of two liquids, the more volatile will boil at a lower temperature, and it will evaporate faster when both liquids are at the same temperature. The volatility of petroleum products can be evaluated by tests for Flash Point, Vapour Pressure, Distillation, and Evaporation Rate.

volt

volt (V) is the SI derived unit of electric potential. One volt is the difference of potential between two points of an electical conductor when a current of 1 ampere flowing between those points dissipates a power of 1 watt. It is named after the Italian physicist Alessandro Volta (1745-1827).

volume fraction

volume fraction (*ö*) is defined as where V_j is the volume of the specified component and the V_i are the volumes of all the components of a mixture prior to mixing.

volume solids

the volume of pigment plus binder divided by the total volume, expressed as a percent. High volume solids mean a thicker dry film, improved hiding, and high durability.

vulcanisation

nonreversible chemical reaction involving sulphur or other suitable agent wherein cross-links are formed between molecular chains in rubber materials.

wafer

small, thin, flat circular disk of a semiconducting material, such as pure silicon, that is masked, oxide-coated, doped, and otherwise processed for ultimate seperation into numerous individual electronic devices or for packaging as in integrated circuit.

wall effect

the consequence of the looser packing density near the walls of the rigid HPLC column. Mobile phase has a tendency to flow slightly faster near the wall

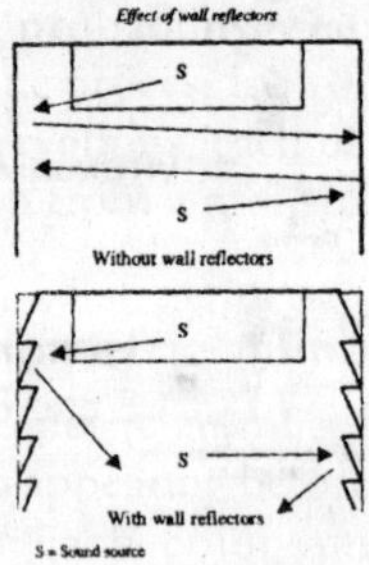

because of the decreased permeability The solute molecules that happen to be near the wall are carried along faster than the average of the solute band and. consequently. band spreading results.

wapdeg

a computer software code used to analyse drip shield and waste package degradation.

wash

term used in the southwest for a broad, gravelly, dry bed of an ephemeral stream, generally in the bottom of a canyon.

washability

ease with which washing will remove dirt from the paint's surface without causing damage.

waste form

a generic term that refers to the different types of radioactive wastes.

waste package

a sealed container containing waste that is ready for emplacement.

The waste package includes the waste form and any containers, spacing structures or baskets, and other absorbent materials immediately surrounding an individual waste container placed internally to the container or attached to the outer surface of the disposal container.

■ waste package remediation system

a repair facility for disposal containers and waste packages that have failed the weld inspection processes, that are defective or abnormal, or that have been selected for retrieval from the repository for performance confirmation examinations.

■ water hardness

a measure of the amount of metallic salts found in water. Hard water can inhibit the action of some surfactants and reduce the effectiveness of the cleaning process.

■ water table

1. the upper limit of the portion of the ground wholly saturated with water.
2. the upper surface of a zone of saturation above which the majority of pore spaces and fractures are less than 100 percent saturated with water most of the time (unsaturated zone) and below which the opposite is true (saturated zone).

■ waterbased

this is a term Nikwax uses to indicate that the solvent for the active ingredients in Nikwax products is water and not the environmentally dangerous chemicals used in other waterproofing products. A waterbased product will deliver the treatment along the same path that water travels into a boot or garment, making it a highly effective solvent.

■ watt

watt (W) is the SI derived unit of power. One watt is a power of one joule per second (W = J/s). It is named after the Scottish engineer James Watt (1736-1819).

■ wavelength

the distance beween neighbouring peaks of a wave of electromagnetic radiation.

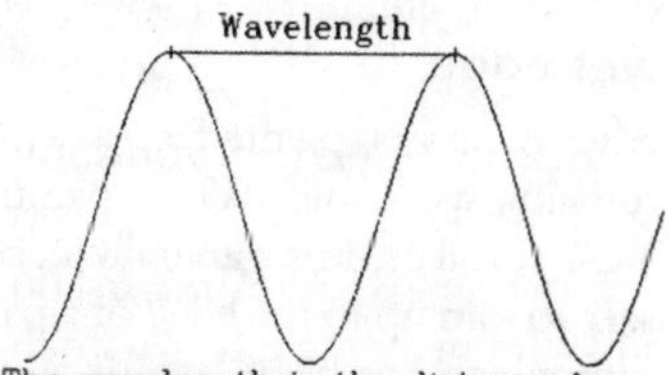

The wavelength is the distance from crest to crest on the wave

■ WAX

weak anion exchanger. Ionisable groups such as primary. Secondary or tertiary amino groups on a packing are considered to be weak.

■ WCX

weak cation exchanger. Carboxylic groups on a packing are typical of a weak cation exchanger.

■ weight per gallon

the weight per gallon of any liquid is determined by multiplying the weight of a gallon of distilled water (8.33 lbs) by the specific gravity of the liquid.

■ **welded tuff**

a tuff that was deposited under conditions where the particles making up the rock were heated sufficiently to cohere. In contrast to nonwelded tuff, welded tuff is considered to be denser, less porous, and more likely to be fractured (which increases permeability).

■ **wet edge**

edge of a wet painted area which remains workable. When painting large surfaces, it is generally necessary to join up to the edge of a paint film which has been left for an appreciable time; when this can be done by blending this edge with free working paint without any lap showing, the film is said to present a wet edge.

■ **wetting agent**

a substance that increases the spreading of a liquid by reducing the tension between two surfaces.

■ **white blood cell**

any of the blood cells that are colourless, lack haemoglobin, and include the lymphocytes, monocytes, neutrophils, eoinophils, and basophils.

■ **whole blood**

a living tissue comprised of red blood cells, white blood cells, and platelets suspended in plasma, that circulates through the body carrying nourishment, electrolytes, hormones, vitamins, antibodies, heat, and oxygen to the body's tissues.

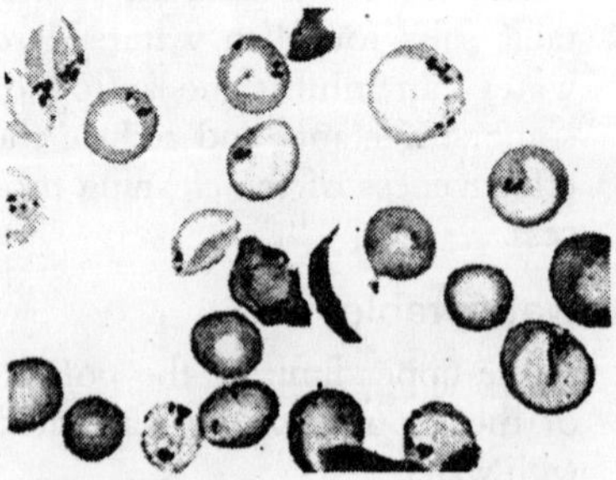

■ **wool wax**

see **lanolin.**

■ **xanthan gum**

also known as corn sugar gum, this is a polysaccharide produced from bacteria (Xanthomonas campestris) fermented with a carbohydrate. It is used in food and cosmetics as a thickener, emulsifier and stabiliser. It has no known toxicity.

■ yield strength

the stress required to produce a very slight yet specified amount of plastic strain.

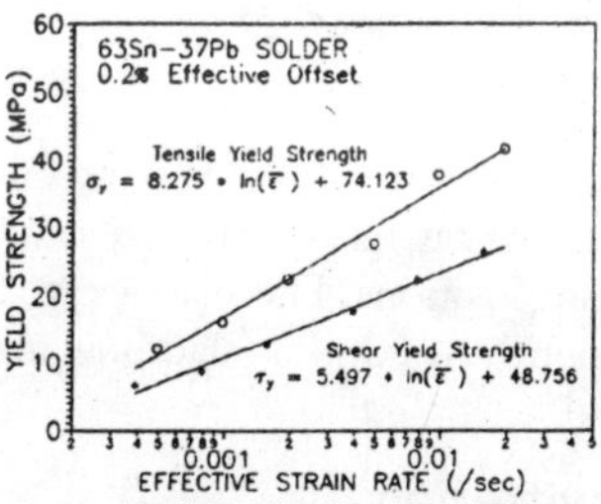

Tensile and shear yield strength data as a function of effective strain rate with corresponding logarithmic fits of the data.

■ young's modulus

see modulus of elasticity the

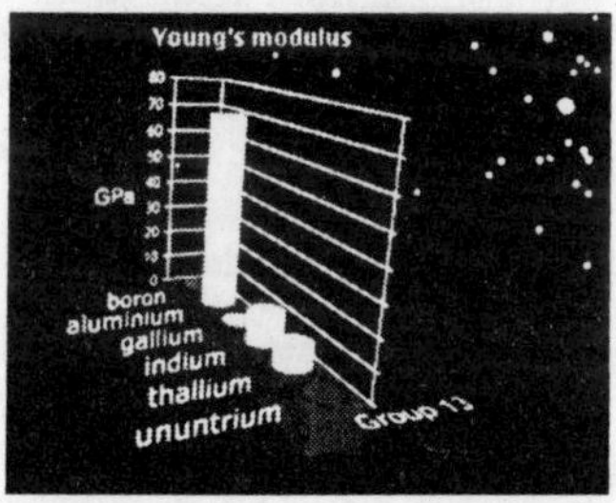

ratio of stress to strain when deformation is totally elastic.

■ Zeeman effect

Zeeman effect is the splitting of the lines in a spectrum when the source of the spectrum is exposed to a magnetic field. It was discovered in 1896 by Pieter Zeeman (1865-1943).

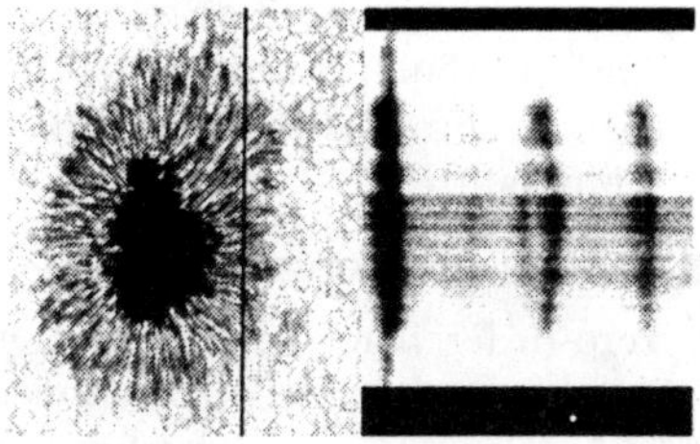

■ Z_{eff}

the effective nuclear charge felt by an electron. The effective charge depends on how well the other electrons in an atom are able to shield (or cancel out) the positive nuclear charge. For example, if the two 1s electrons in Li were perfectly shielding then the Z_{eff} felt for the outermost electron would be 1 unit.

■ zeolite

zeolite is a natural or synthetic hydrated aluminosilicate with an open three-dimensional crystal structure, in which water molecules are held in cavites in the latice. The water can be driven off by heating and the zeolite can then absorb other molecules of suitable size. Zeolites are used for separating mixtures by selective absorption.

■ **zeotrope**

zeotrope is a liquid mixture that shows no maximum or minimum when vapour pressure is plotted against composition at constant temperature.

■ **zero-order reaction**

zero-order reaction is a reaction for which the rate of reaction is independent of the concentration of reactants.

■ **zeta potential**

1. zeta potential (*æ*) is the electric potential at the surface of a colloidal particle relative to the potential in the bulk medium at a long distance. Also called electrokinetic potential.

2. zeta potential (*æ*) is the potential across the interface of all solids and liquids. Specifically, the potential across the diffuse layer of ions surrounding a charged colloidal particle, which is largely responsible for colloidal stability. Also called electrokinetic potential.

■ **Ziegler process**

Ziegler process is an industrial process for the manufacture of high-density polyethene using catalysts of titanium(IV) chloride ($TiCl_4$) and aluminium alkyls (e.g. triethylaluminium, $Al(C_2H_5)_3$). The process was introduced in 1953 by the German chemist Karl Ziegler (1898-1973). It allowed the manufacture of polythene at lower temperatures (about 60 °C) and pressures (about 1 atm) than used in the original process.

■ **zircaloy**

a family of alloys of zirconium that may have any of several compositions. These alloys are frequently used as a cladding material.

■ **zwitterion**

zwitterion is an ion that has a positive and negative charge on the same group of atoms. Zwitterions can be formed from compounds that contain both acid groups and basic groups in their molecules. It is also called ampholyte ion.

■ **zyme**

a fermenting substance.

Other Books on

LOTUS ILLUSTRATED DICTIONARIES

33. Geology	95/-
34. Genetic Engineering	95/-
35. Health & Nutrition	95/-
36. History	95/-
37. Import and Export	95/-
38. Information System Management	95/-
39. Internet	95/-
40. IT	95/-
41. Inorganic Chemistry	95/-
42. Law	95/-
43. Library & Information Science	95/-
44. Literature	95/-
45. Management	95/-
46. Mathematics	95/-
47. Marketing & Sales	95/-
48. Mass Communication	95/-
49. Mechanical Engg.	95/-
50. Medical	95/-
51. Music	95/-
52. Organic Chemistry	95/-
53. Philosophy	95/-
54. Physical Education	95/-
55. Physics	95/-
56. Psychology	95/-
57. Science	95/-
58. Sex	95/-
59. Sociology	95/-
60. Sports	95/-
61. Textile	95/-
62. Zoology	95/-